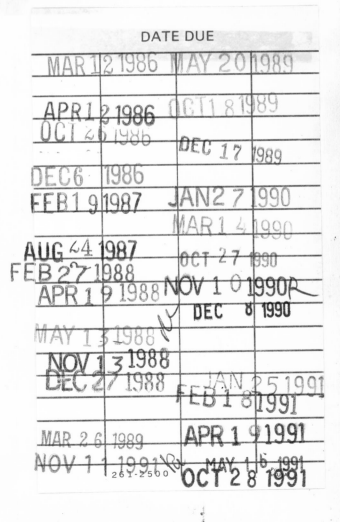

EXPERIMENTING
WITH
PLANTS

EXPERIMENTING WITH PLANTS

Projects for Home, Garden, and Classroom

JOEL BELLER

Francis Lewis High School
Flushing, New York

ARCO PUBLISHING, INC.
NEW YORK

Published by Arco Publishing, Inc.
215 Park Avenue South, New York, N.Y. 10003

Library of Congress Cataloging-in-Publication Data

Beller, Joel.
  Experimenting with plants.

  Bibliography: p.
  Includes index.
  1. Gardening—Experiments.   2. Botany—
Experiments.
I. Title.
SB454.3.E95B45   1985   580'.7'8   85-9184
ISBN 0-668-05989-3   (cloth edition)
ISBN 0-668-05991-5   (paperback edition)

Printed in the United States of America

10 9 8 7 6 5 4 3 2 1

CONTENTS

LIST OF ILLUSTRATIONS

1   ARE PLANTS
FOR YOU?

Congratulations! Reading this book means you are at least considering a science project or experiment. Perhaps you are convinced you will do something.

No matter what your reason for choosing to do a science project, I am going to do my best to convince you to do a botanical project or experiment. Just look at the smiling, happy face of the young lady with her plant project in Figure 1.1. She was a semifinalist in

Figure 1.1. A student with her science project

1

the 1984 Westinghouse Science Talent Search. This could be you in the near future.

ANIMALS ARE FUN BUT . . .

Most people believe that animals are much more exciting than plants. After all, animals move. Well, so do plants! Most plants move very slowly. They grow tall and bend toward the light. Others attach themselves to fixed objects by wrapping around a pole, fence, or even another plant (see Fig. 1.2). The most spectacular "mover" is the Venus flytrap. It can entrap an insect by quickly closing two adjacent leaves if the insect touches the hairlike trigger on this living trap. Plants certainly do move!

Most animals require feeding, cleaning, and care at least once a day. Their wastes are often smelly and can be annoying to your

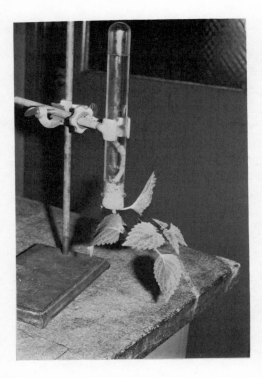

Figure 1.2. A coleus, stem bending upward. This is an illustration of plant movement.

parents. If you go away on a weekend trip, you may have to ask a friend or neighbor to babysit for your experimental animals. Green plants, on the other hand, may require only a heavy watering and plastic wrap to prevent excessive water loss. "Gentle neglect" is best for the welfare of most plants.

Animals also have the annoying habit of becoming pets rather than an essential part of your project like a thermometer or motor. Plants rarely develop personalities the way animals do. Young people, and older ones, too, often get emotionally upset when an animal dies. There is less chance this will happen with a plant.

DON'T SELL PLANTS SHORT

Plants come in all sizes, shapes, and colors. Some plants need lots of water, while others thrive under desert conditions. No matter where you live you can grow plants of some sort. You can grow plants inside your apartment or house or out-of-doors if you have a terrace or yard in the city as well as in the suburbs or country. You would probably be amazed by the number of different species of plants that you can grow at home or in school. With a little effort, for example, it's possible to create your own small tropical rain forest, and to grow even delicate orchids at home.

Most science competitions, from local science fairs to the Westinghouse Science Talent Search, prohibit projects or experiments involving vertebrate animals. By working with plants, you needn't be concerned about this restriction. Plant projects are accepted in all science competitions and expositions.

Another nice thing about plants is that they never complain if you neglect them. True, they may die or grow poorly, but complain? Never! Are you going to design and carry out a project to prove me wrong on this point? If so, good for you!

When you are finished with your plant project, you might be able to eat it—that is, if your project involves carrots, tomato plants, or some other fruit or vegetable. I have yet to hear of a student who ate the amoebas when the experiment was concluded.

When it is Mother's Day or your mom's birthday, part of your plant project might make an ideal gift for her. Add some colorful wrapping paper, a ribbon, and a nice card to a geranium plant or an African violet and you have a wonderful present—a very special one, cultivated and grown by you.

I have listed a number of advantages to using plants in science projects. How many can you think of that I missed?

2   GETTING STARTED

When you finish this book you will be able to design and most likely carry out a plant project that is scientific and fun to do. Your thought might be, "Aren't all projects scientific?" The answer is, "No!" Doing an experiment or project scientifically means you are following a procedure acceptable to scientists. Although there is more than one "scientific method," we shall concentrate on one method, namely, the controlled experiment.

PREPROJECT PLANNING

When you do an experiment or project you are really asking a question—for example, will snake plants grow taller if I add an enzyme, super oxide dismutase (SOD), to their water? The scientific term for the question is *hypothesis*. How you answer the question is the procedure.

The scientific way to answer the question of super oxide dismutase's effect on snake plant growth is to first set up two groups of snake plants. One will be the experimental group and the other the control group. Both groups must be as alike as possible in all respects. The plants in both groups must be the same in size, age, state of health; have the same soil and pot type; be exposed to the same light and temperature, etc. The only difference (*variable*) is that the experimental group will be watered with a SOD solution and the control group will receive plain water. Your results (*data*) will be the amount of new growth (in centimeters or millimeters) that occurred in both groups after a period (e.g., a month, six weeks) of experimentation. If the experimental groups show significantly more growth, then your experiment's conclusion is positive. A negative conclusion would be reached if your data show no difference in growth between both groups.

Two things to keep in mind when doing a controlled experiment are (1) avoiding favoritism, and (2) sufficient numbers. By "favoritism" I mean giving a little extra water to the experimental plants or placing them where they get just a bit more light than the controls. If you favor one group of plants, your experiment is no longer a controlled experiment and is unscientific.

You must have sufficient numbers of plants in both groups in order for your results to be meaningful. Let's imagine you had only one plant in each group and the snake plant receiving the SOD died. Your conclusion would be that SOD is toxic to snake plants. Suppose you used 15 plants in each group and one experimental plant died but the other 14 plants grew an average of 12 centimeters (cm) taller than the 15 plants in the control group. Your conclusion, obviously, would be different.

DO YOUR OWN THING

As you read more chapters, you will discover many possible project formats involving different plants. What you decide to do depends on many factors and variables. For example, you might love the sight and smell of roses. For you, roses might be a good choice for a project. On the other hand, if you must do your project in the winter and time is limited, your project might center on forcing freesia bulbs to flower quickly. Lettuce or radishes are possible choices if you need to start a project in the early spring and finish it before the hot weather arrives. Spider plants (see Fig. 2.1) can be the basis of your experiment if you live in an apartment where space and light are limited.

After you decide what plant you want to work with, decide what you want to do. You may have heard, for example, that peony seeds need to be kept cold for several months before they will germinate. You may have discovered in your reading that gibberellic acid affects the length of time that seeds must remain dormant before germinating as well as having other effects such as speeding up growth. These facts may spark your curiosity and seem worth investigating further. However, just saying, "I want to do something

Figure 2.1. Spider plants can be the subject of your project.

with peony seeds and gibberellic acid!'' is not enough. You must be more specific. A hypothesis for a possible experiment could be, ''What is the effect of gibberellic acid on the refrigeration time that peony seeds need for germination?''

Suppose you decide to try this project; how will you go about doing it? In the rest of this chapter, I will go through the procedure step by step. Bear in mind that this is just a model experiment and that the steps you take with your own plant project will probably be different.

SEARCH, READ, RECORD

No matter what project you choose, your first stop is the library. Go to the subject card file. Here you can find out what others have written on the topics relevant to your project. The first subject areas to search for our model experiment are plant growth substances, gibberellic acid, peony plants, seeds, horticulture, and dormancy.

Go to the best library for your purposes that you can. For example, the reference room of your public library might have newer books and more of them than your school library. Your science teacher might suggest and even help you to get technical books from a nearby college or university library.

Your library visit (see Fig. 2.2) is not complete unless you examine the scientific encyclopedias. Find out what they say about the subjects you are interested in. Also, search any subject titles they suggest you explore.

Your next stop is the reference room of the library. Ask for the *General Science Index*. This *Index* comes out ten times each year and lists, by subject and author, all the articles in science journals and magazines (for example, *The American Journal of Botany* and *Scientific American*) suitable for teenage and young adult readers. This *Index* is a better source of current science articles than the

Figure 2.2. The library is the place to begin your project.

Reader's Guide To Periodical Literature. The *Guide* lists articles that have appeared in all sorts of journals, not only journals that specialize in science. Naturally, if your library doesn't have the *General Science Index,* use the *Reader's Guide to Periodical Literature.* Both the *Index* and the *Guide* list the articles by subject. They tell you the title and author of the article, and the name and volume of the magazine or journal where each article can be found. Look for articles that might have information related to your experiment.

Another helpful index of research papers and articles in science is the *Science Citation Index, Abridged Edition.* This literary aid helps you locate all current articles published by any specific scientific author. The *Science Citation Index* also has a subject index based on key words and terms. Unfortunately, not all libraries have this Index.

Highly technical papers from professional journals such as the *American Peony Society Bulletin* are not listed in either the *Index* or the *Guide.* For these, you will have to go to a special library in a large city or to a nearby college or university library. Many colleges will allow you to use their facilities if you bring a letter from your school.

Don't overlook the *New York Times Index* if it is available. Many worthwhile science articles can be found in the *New York Times.* The articles are stored on microfilm, and you can read them using the library's special viewer. Bring a pen and notebook or large index cards in order to record any valuable information that you find. If you can, use the copying machine to duplicate important articles. It will be well worth the money!

If your library is really well equipped, you can make a computer search of a data bank for the same information you may find in the *General Science Index.* The librarian will help you do this. Some libraries perform this service without charge; others charge a fee. Another possibility is that your school or the central school office may have access to a data bank. If you have a personal computer and a modem—a piece of equipment that connects your computer to the telephone—you can access data banks such as The Source. Data banks charge for the time you spend searching their files, however, and you may have to pay a monthly fee in addition to what you pay each time you use the service.

If your search of the literature is successful, you will discover the following:

Fact 1: Peony seeds should be planted in the spring after being refrigerated for eight weeks at 2° to 5°C.

This means you must keep them at the bottom of the refrigerator for two months. Will your parents be annoyed by seeing the seeds in the refrigerator day after day? Will they throw them out after a few weeks? Did your science teacher say your project must be completed in six weeks? If the answer to any of these questions is "Yes," get another project!

Fact 2: Peony seeds germinate and grow best in the colder parts of the United States.

If you live in a warm part of the country such as Florida, it would be wise to select another project.

Fact 3: Some peony seeds will germinate two years after the period of refrigeration. Sometimes, even longer refrigeration time is needed.

Thus, the more seeds you use in your experiment, the better your chances that some will germinate in a reasonable length of time after refrigeration. Two months is not unreasonable.

Fact 4: The best way to propagate peonies is not to use seeds but to replant pieces of roots taken from an old peony plant.

Growing peonies from seed takes care, skill, and most likely a bit of luck! Is this project for you?

Fact 5: Peonies need plenty of room to grow. They should be planted 3 meters (m) away from each other.

The large amount of space these plants require might be a problem if you want to go beyond the germination stage and you live in a 20-story apartment building in the middle of a city.

Fact 6: Peonies are subject to few diseases and are attacked by only a few pests such as nematodes (soil worms). Hurrah! Finally, a fact that favors doing an experiment with peonies.

Fact 7: Gibberellic acid causes enzymes in seeds to break down stored food to a usable form. Thus, germination can occur and the embryo grows. Perhaps the action of gibberellic acid will shorten the cold dormant period normally required for peony seed germination.

Read up thoroughly on the topic you plan to investigate and take notes on what you have read. Then, go over the information carefully and ask yourself: Is this project for me?

Let's suppose, after reviewing the facts, that you decide this peony seed and gibberellic acid project is the project for you. Let's go through the planning for the project.

- Outline what you intend to do.

- Prepare a list of materials and supplies needed for the experiment.

- Restate your hypothesis. In this case: "Will gibberellic acid shorten the refrigeration time needed for peony seeds to germinate?"

For a start, let's try four different concentrations of gibberellic acid: 1 part per 100 of solution, 1 part per 1,000, 1 part per 10,000, and 1 part per 100,000. Don't forget that you must have a control group for each gibberellic acid solution (see Table 2.1). You may wonder why a dilution of 1 part per 100,000 of solution is used. The reason is that plant hormones are normally present in most plants in amounts this small. Shortly, you will learn how to prepare these dilute solutions.

The next item is peony seeds. How many should you use? The error made most frequently by beginning scientists is that they tend to work with too few organisms in their experimental and control groups. Large numbers are needed in each group if the groups are to be compared in a meaningful way. Since this is your first scientific experiment, use a minimum of 25 seeds per group. The more, the better (see Fig. 2.3)!

Keeping in mind that the normal refrigeration period is eight weeks, let's try seeds treated with gibberellic acid for different periods varying from zero to eight weeks. Let's also attempt to germinate peony seeds that were not refrigerated at all. We will refrigerate some seeds for only two weeks, another group for four weeks, a third group for six weeks. Our final batch of seeds will be refrigerated for the full eight weeks.

This last batch is very important, especially if no seeds in any group germinate. This can happen if you were supplied with old

Figure 2.3. Many plants must be used in a controlled experiment.

seeds or improperly stored seeds. Another possibility is that the fault is yours! Perhaps your refrigerator is too hot or too cold, not in the 2°-to-5° C range that peony seeds require. This is why an accurate Celsius thermometer is so important in this experiment. Take frequent readings to be sure your seeds are being chilled within the correct temperature range.

Looking at Table 2.1, you can see that anyone doing this experiment needs 25 groups of 25 seeds each, or a total of 625 seeds. This may seem like a lot to you, but a senior scientist doing this particular experiment would use a minimum of 100 seeds per group. That means at least 2,500 seeds would be used.

SERIAL DILUTION

Let's turn our attention now to the gibberellic acid solutions. A number of biological supply houses can supply you with a 1% solution of gibberellic acid. Another way of expressing 1% is 1 part gibberellic acid to 99 parts water. For our purposes, let's think of it

Table 2.1
GIBBERELLIC ACID

Seed Group	Solution Concentration	Refrigeration Time (in weeks)
1 (control)	0 (pure water)	0
2	1:100	0
3	1:1,000	0
4	1:10,000	0
5	1:100,000	0
6 (control)	0 (pure water)	2
7	1:100	2
8	1:1,000	2
9	1:10,000	2
10	1:100,000	2
11 (control)	0 (pure water)	4
12	1:100	4
13	1:1,000	4
14	1:10,000	4
15	1:100,000	4
16 (control)	0 (pure water)	6
17	1:100	6
18	1:1,000	6
19	1:10,000	6
20	1:100,000	6
21 (control)	0 (pure water)	8
22	1:100	8
23	1:1,000	8
24	1:10,000	8
25	1:100,000	8

as being 1 part gibberellic acid to 100 parts solution. To get the other solutions, you have to prepare a series of dilutions starting with the 1% solution. You will need a 10-milliliter (ml) graduated cylinder, a 100-ml graduated cylinder, several flasks or test tubes with corks to store the solutions, and a marking pencil and labels (see Fig. 2.4).

Preparing pure water should be no problem. To rule out any

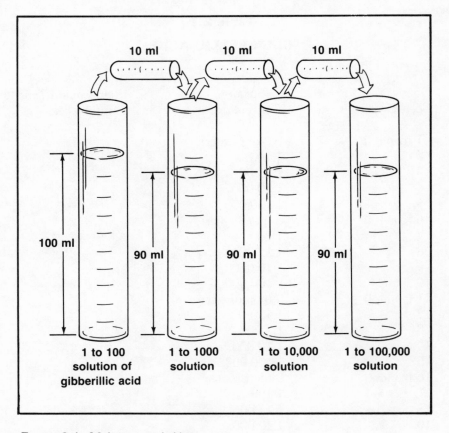

Figure 2.4. Making serial dilutions

chance of contamination, use demineralized water. The 1 part per 100 is also easy since that is the 1% solution you purchased from the biological supply house. To prepare the 1 part of gibberellic acid in 1,000 parts of solution is easier than it sounds.

Using the 10-ml graduated cylinder, carefully measure out 10 ml of the 1% solution. Make sure the meniscus (the bottom of the water level) is on the 10-ml line. Pour the 10 ml into the 100-ml graduated cylinder and add enough water to make 100 ml of solution. The strength of this solution is 1 part gibberellic acid in 1,000 parts of solution. Pour this into a labeled container and save.

You have just learned the basic technique for diluting solutions. You are now ready to make up the 1:10,000 and the 1:100,000 solutions.

Remember to mark and label all flasks or test tubes. Also, wash and dry the 10-ml graduated cylinder.

SEED PREPARATION

Your next task is to soak the first five groups of 25 seeds for four hours in plain water. Soak groups 6 to 10 for four hours in the 1 to 100 solution. Treat the remaining groups as indicated in Table 2.1.

After soaking, place the seeds in groups 6 to 25 in individual plastic envelopes. Each group goes in a separate, marked envelope. Put the envelopes in your refrigerator in the place where the temperature will always be between 2° and 5° C. Temperature and time are very important during this phase of the experiment. Keep a thermometer there and check the temperature from time to time. Don't forget to mark on your calendar the dates for removing the various seed groups from the refrigerator. The last phase of the experiment is to observe which seeds will germinate.

GERMINATION

Petri dishes are convenient and inexpensive items to use as germination chambers. Line the bottom of the dishes with moist absorbent or blotting paper. Remember, there is a big difference between *moist* and *soaked*! If the seeds get too wet, they will drown and not germinate. Seeds must be kept moist at all times when you are trying to germinate them. No moisture is just as bad as too much! You will need 25 germination chambers, one for each group. Write the group numbers on the bottom of the petri dish first. Then place all 25 seeds of each group in the same germination chamber. Figure 2.5 shows you what one looks like with the 25 seeds inside.

Seal each labeled chamber and add the date germination was started to the information on the bottom of the germination dish. Germinating seeds should be maintained in a cool environment where the temperature is between 15° and 21° C. Check your seeds daily, making sure they are moist; if necessary, add a bit of water. Germinate your seeds for a month or more. Keep careful records. One way is to write your observations in a log or diary. Photographs are a good way to enhance your observations.

**Moist blotting paper
(on bottom of dish)**

Peony seed

Petri dish cover

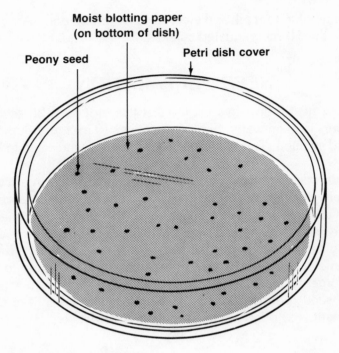

Figure 2.5. Germination chamber

CONCLUDING STATEMENTS

At the conclusion of the experiment, let's imagine that your data look like that presented in Table 2.2.

Notice that in no group did all 25 seeds germinate. This is to be expected because rarely are all the seeds viable. If 85% of the seeds in a batch germinate, you are doing about average. Usually, the fresher the seeds, the higher the germination percentage.

One conclusion is that the effect of gibberellic acid in concentrations of 1 part per 100 and in 1 part per 1,000 seems to be just about the same. It is also apparent that 25 seeds per group is really too small a sample. At this point, it would be wise to refine the experiment and work only with these two concentrations of solution. Use 100 or more seeds in each group. Your second experiment might help reveal the most effective concentration.

In chapter 10, you will see how an experiment is written up for others to read.

Table 2.2

EXPERIMENTAL AND CONTROL GROUPS

Refrigeration Time (in weeks)	Concentration of Gibberellic Acid				
	Zero	1:100	1:1,000	1:10,000	1:100,000
0	*(1) †0	(2) 0	(3) 0	(4) 0	(5) 0
2	(6) 0	(7) 1	(8) 0	(9) 0	(10) 0
4	(11) 0	(12) 8	(13) 9	(14) 0	(15) 0
6	(16) 2	(17) 13	(18) 11	(19) 6	(20) 1
8	(21) 21	(22) 23	(23) 22	(24) 23	(25) 22

*( ) = Group number
† Number of seeds that have germinated is indicated by figures in second row of each time block.

SPIN-OFF EXPERIMENTS

You may realize after completing this experiment that working with peony seeds is fun. Moreover, you would like to do a follow-up project along similar lines. What can you do? Here are just a few ideas for experiments that spin-off from the peony seed project:

● Is gibberellic acid more effective as a paste than as a solution?

● How do the experimental peony plants compare with the control plants in size, shape, number of flowers, resistance to disease, etc?

● What is the effect of various concentrations of N-diethyl-amino succinic acid on peony seeds?

Many youngsters in search of an experiment end up doing a spin-off to something they read about or saw. So, visiting science fairs and competitions can be helpful because you may see a project that seems exciting and arouses your creative urge to take that project a step further. Yours will be similar but different. Each experiment is a building block of scientific knowledge, and practically every researcher uses some other scientist's experiment for developing a more elaborate and sophisticated experiment. This is how scientific knowledge is gained. Don't be afraid to spin off and then take off on your own!

3   SEEDS TO SEEDLINGS

No matter what plant project you choose, one thing is for sure: You must have plants! You have to start your plants from seeds, or begin with grown plants, or use pieces of plants. This chapter covers starting plants from seeds, while later chapters tell you how to use small pieces of grown plants for growing more. Starting your plants from seeds is fun and rewarding. You will get a feeling of accomplishment when you discover a white stem sprouting where there was only soil the last time you looked.

The plans discussed in this chapter work for most seeds. For unusual plants, follow the directions on the seed package or consult one of the horticulture books listed in the Bibliography at the end of this book.

SEED SELECTION

First you must decide what kinds of seeds you want for your project. Some seeds are easy to work with, and their germination rate is high because these seeds are not too "fussy." In this category are some vegetables that have a high success rate, such as beets, lettuce, radishes, cucumbers, and tomatoes (see Fig. 3.1). Don't try to work with celery seeds, as they are hard for beginners to grow at home. Avoid brussel sprouts and cauliflower unless you are an experienced vegetable gardener.

Flowers such as carnations, coleus, and pansies have seeds that are good for beginning gardeners; so do marigolds and zinnias.

You may have some seeds in your home that you can plant, for example, unroasted whole coffee beans, or grapefruit, orange, and lemon seeds. Unpasteurized date seeds (sold in many health

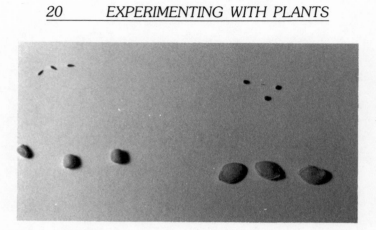

Figure 3.1. Lettuce seeds (upper left), cabbage seeds (upper right), corn seeds (lower left), squash seeds (lower right). The size of the seeds determines the planting depth. Lettuce seeds can be scattered on the surface of loose soil. Cabbage seeds need to be planted 0.5 cm deep. Corn and squash are planted 3 cm deep.

food stores) and grape seeds also will yield plants. Avocado seeds work well: Just remove the pit from a ripe avocado, then wash it. Bury three-quarters of the pit in the pot and keep it watered. In less than a month you will have a plant that keeps getting taller and taller unless you force it to grow side branches by pinching off the growing tip after the plant is about 10 cm tall.

SEED NEEDS

Most seeds contain an embryo plant and some stored food to provide nourishment until the baby plant can feed itself by photosynthesis after germination. During the germination period, seeds must be kept constantly moist and warm. The medium the seeds are planted in should be lightweight and porous. The tender roots emerging from germinating seeds can't push their way through heavy dirt, and a porous medium holds lots of air and allows water to flow easily. Notice that nothing was said about providing minerals. This is because the stored food in the seed meets all the embryo's nutritional needs. After germination, when the second pair of leaves has appeared, transplanting to a heavier and more nutritious soil is necessary.

GROWTH CHAMBERS

There are many kinds of starting containers, some made of clay, others of plastic.

Jiffy Pots® are made from peat moss and vary in size from 5 to 8 cm in diameter. Some Jiffy Pots® are square; others are round. Peat is a wonderful material with which to make seed containers. You can plant the young seedling, pot and all! The roots will grow right through the sides and bottom of the pot.

Fill the Jiffy Pot® with your starting mix, bury the seed to its proper depth, set the pot in a container of water, and you are in business.

You can also buy Jiffy Peat Pellets® that contain a ready-made mix of soil, humus, and nutrients. These are sold in a dry, compressed form. When placed in water they expand into a 5-cm seed germination chamber. Using a pencil, make holes in the top and add the seeds to their recommended depth.

Another type of germination container is a small plastic pot with drainage holes in the bottom (see Fig. 3.2). These usually come in a set of 20 or more, complete with a plastic watering tray into which they fit snugly. Just fill with the starting mix of your choice, add seeds, and water. Keep the temperature between 21° and 27° C and you are all set. That temperature range favors seed germination.

You can also make your own germination chambers, which will save you money. Use empty containers from frozen orange or grapefruit juice, yogurt, frozen fruit, cream cheese, etc. Don't forget that your homemade seed containers must sit in a waterproof tray of some sort so that the germinating seeds have a constant supply of moisture. All you have to do now is add the seeds.

STARTING MIX

Commercial "starting mix" is probably the best and certainly the easiest medium to use. You can also make your own mix. Use one part milled peat moss (a dried bog plant that is light and has excellent water-holding qualities; the milling process makes the moss light and fluffy so that it holds air as well as water), one part vermiculite (light, porous, made from the mineral mica), and one part perlite (made from ground-up volcanic rock).

Figure 3.2. Plastic germination pots in their watering tray. The seedlings are lettuce plants.

SEED DEPTH

Always sprinkle lettuce seeds on the surface of the planting mix, whether you are starting them in Jiffy Pots® or in your garden. Plant carrots about 1 cm below the surface of the planting mix; cucumbers, about 3 cm. If the seed package has no instructions, multiply the diameter of the seeds by 1.5 and plant the seeds at this depth.

WARNING

Whenever you handle seeds, chemicals, fertilizers, insecticides, *keep your hands away from your mouth!* Practice cleanliness and wash your hands thoroughly after any experimentation.

GERMINATION TEST

Seeds kept in a tightly sealed jar in a cool place will often germinate, even if several years old. Eventually, however, seeds will die. Here is a simple way to discover if your seeds will germinate:

1. Moisten a paper towel thoroughly.

2. Cut to fit inside a saucer.

3. Sprinkle 10 seeds onto the moist towel.

4. Slip the saucer, towel, and seeds into a plastic bag and seal the bag with a tie.

5. Place the bag in a warm spot such as on top of the refrigerator. The temperature should be around 23° C.

6. Examine the seed package and note how many days must pass before the seeds germinate. On the day that the seeds are supposed to germinate, remove the bag and look to see if the first small roots have broken through the seed coats. Wait three more days. Then, count the number of seeds that have germinated.

7. If 8 seeds germinated, then the germination rate is 80%; if only 3 seeds have germinated, the germination rate is 30%. If the germination rate for the seeds you are using is 40% or less, purchase new seeds: your chances of getting most of the seeds to germinate are poor, especially if you intend to plant the seeds out-of-doors.

8. Don't try to save the seedlings from the germination test. Most will die after planting because their tiny, delicate roots had been seriously damaged.

Keep this method in mind if your project calls for applying a chemical solution that may affect the speed of germination or if you have to measure root growth in seedlings.

PORTABLE GREENHOUSE

A greenhouse is a good place to grow your seedlings, since you can easily regulate the temperature and humidity. It is possible to buy a small greenhouse, but you can make one for far less money and also experience the satisfaction of building something useful by yourself. The basic parts are some supports to serve as a frame (wooden ones are fine), and a clear plastic cover. Figure 3.3 shows a diagram of a greenhouse for a seed tray that measures 20 by 30 cm.

Plastic wrap won't do because it's too thin. Thicker plastic is easier to work with because it can stand rougher handling. Use plastic at least 4 millimeters (mm) thick, which you can buy at a hardware store; this can be the plastic sold to make inexpensive storm windows.

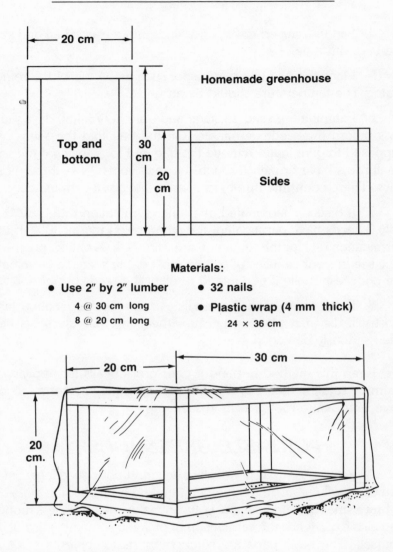

Figure 3.3. Diagram and dimensions for making a homemade greenhouse

HOW MANY SEEDS DO YOU PLANT?

It is always wise to plant more seeds than the number of adult plants you want. This holds true for indoor and outdoor planting, since for one reason or another you can be sure that some of your seeds won't germinate. Plant three seeds in each pot. If they all germinate, thin out the pot by pinching off the tops of the least healthy plants. Pinching off the tops is a better method than pulling out the unwanted seedlings. The seedling you select to live may have its root system damaged when nearby seedlings are removed.

If your seedling house plants are to survive and prosper, careful attention must be given to soil, water, light, temperature, and air quality. The quality of the air is determined by its temperature, relative humidity, and pollutants. Let's examine each of these essentials, one at a time.

Dirt Science

After you select the proper pots for your plants, you will have to fill them with potting soil. The three main ingredients in soil are sand, clay, and humus. The sand allows water to drain through the soil easily, while the clay binds together the sand and humus. The humus, which is dead plant and animal matter, helps to retain the water and supplies needed minerals.

You can make your own humus by a process called *composting*. A pile of plant wastes including leaves, grass clippings, and kitchen wastes will be decomposed into simpler form by soil microorganisms that include bacteria, molds, and other fungi. The plant wastes are usually kept in a bin made of wire, wood, or plastic. The compost pile has to be kept moist and aerated if the soil microorganisms are to reproduce and do their job. If correctly moisturized and aerated, there won't be any odor from your compost pile, but it *can* be unsightly, so select a back corner of your garden (if you have one) as the site of your compost pile. If you don't have access to waste products, gardening establishments sell compost-starting tablets or flakes, which really are concentrated soil microorganisms in dried form. (It is possible to make compost in small quantities *inside* your house or apartment, but the process can be messy and it is *not* recommended.)

A good basic potting soil can be made by mixing one part builder's sand, one part garden soil, and one part peat moss or humus. If you don't have a garden, you are better off purchasing a bag of potting soil in a garden shop, hardware store, or supermarket.

You can also get special soil mixes for plants with special needs. For example, cactus plants need more sand in their potting mix than other plants. The zebra plant (*Aphelandra*) thrives in soil with two parts humus to one part each of sand and clay. Begonias like a lot of humus and very little clay. These days, many house plants are grown in a mix that contains no soil at all. Because these mixes are easy to use and have a lot of fertilizer, they are good for growing house plants under fluorescent lights. Soilless mixes usually contain vermiculite, perlite, peat moss, a wetting agent, lime, and fertilizer. Jiffy Mix® and Super-Soil® are the names of two of these soilless mixes.

Sterilization

Many experts recommend using sterilized soil. You can either buy it that way or make your own. To sterilize soil, put the soil in a pie plate or baking pan and heat it in the oven for an hour at 350° F. Be aware that the process of baking soil doesn't smell nice. Don't use chemicals to sterilize your soil, because some of them—for example, formaldehyde—are dangerous.

In recent years, other expert gardeners have recommended using *pasteurized* soil instead of sterilized soil. The reason for this trend is that sterilization kills *all* living things in the soil—the helpful bacteria as well as the harmful ones, weed seeds, and viruses. To kill only the harmful organisms by pasteurization, place the soil in an aluminum dish, moisten the soil, then heat it for 20 minutes at 175° to 180° F in the oven. Use an oven thermometer stuck in the soil to make sure you don't exceed 180° F. Check the temperature every five minutes.

Fertilizers Indoors

If the leaves of your plants are yellowish, adding a fertilizer with a high nitrogen content might be the answer. If you want to be sure the flower buds will be healthy, use a fertilizer rich in potassium. Don't buy lawn or garden fertilizer to feed your house plants; outdoor fertilizers dissolve in water very slowly. Your indoor plants need a fertilizer that dissolves almost immediately, although the amount you use depends on how much light they're getting. Plants grown under continuous artificial light never stop and rest. For them the "sun" is always out and there is never a cloudy day. They need a lot of fertilizer. Fertilizing these plants once a week ensures a steady supply of nutrients.

House plants that receive natural light require fertilizer about once a month. Don't concern yourself too much with what the label on the fertilizer package says. Instead, look up your particular plant's fertilizer needs in a good reference source, such as *The New York Times Book of House Plants,* or *All About House Plants* (see Bibliography). Some plants grow quite slowly and may need fertilizer only twice a year.

Adjusting pH

The pH of the soil is as important as water and the minerals present. Most plants grow well when the pH of the soil is close to neutral. (See Bibliography for books on plants with special needs.) The pH scale (see Fig. 3.4) runs from 0 to 14. A pH of 7 is neutral; less than 7 is acid; above 7 is alkaline (basic). The further away from 7 you go, the greater the acidity or alkalinity. Indeed, the difference between two numbers on the pH scale is really a tenfold difference. This means that if a substance's pH is 5, the substance is 10 times more acidic than one whose pH is 6. A substance that has a pH of 4 is 100 times as acidic as a substance whose pH is 6.

To make potting soil more acidic, add one drop of vinegar (pH 3) to the potting soil, mix well, and check the pH. Check after each drop and stop adding vinegar when you reach the desired pH. Increase alkalinity by adding a pinch of bicarbonate of soda; mix well and check the pH before adding another pinch.

ARE YOUR PLANTS ALL WET?

The most common mistake indoor gardeners make is to over-water their plants. This is just as bad as underwatering. If you wet only the surface of the soil, water will never reach the roots. This is the same as if while eating you kept all the food in your mouth and never swallowed any of it. Here are two basic rules:

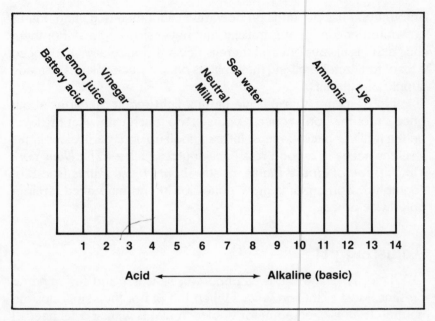

Figure 3.4. pH scale

1. When you water any potted plant, do it *thoroughly*. Stop watering when water leaks out of the drainage holes at the bottom of the pot.

2. Let the soil dry out before watering again. This permits air to reach the roots from the spaces in the soil previously filled with water.

Use common sense:

• Know your plants. Cacti need far less water than ferns. Fatsia does best if its soil is kept constantly moist.

• Use water close to room temperature. Don't shock your house plants by using very cold or very hot water. If you drop cold water on the leaf of an African violet, a permanent spot will form where the drop landed.

• Determine the soil mix. Sandy soil requires more frequent watering than soil with high humus content.

• Water in the morning if possible. As the day progresses, the temperature usually rises and the plant tends to lose water from its leaves. This process is called *transpiration*.

LET THERE BE LIGHT

Once the seedlings break through the surface of the soil, it is important to place them in the light. The more light, the better! Once in the light, chlorophyll production will occur. As soon as the baby plant turns green, it has the ability to manufacture all of its requirements for health and growth, using carbon dioxide from the air and water and minerals from the soil.

Unless you are using artificial lights, your indoor plants will get most of their light through the windows. In the northern hemisphere, the strongest, most direct sunlight enters a room from the south. Light entering from the north may be bright, but the direct rays of the sun will not enter. Easterly windows admit sun during the morning hours, while westerly windows admit sunlight in the afternoon. If you live in an apartment that gets very little light, you may have to use artificial lighting to carry out your projects.

If you must use artificial light, your best choice is fluorescent light bulbs because their light intensity is even, cooler, and cheaper than incandescent bulbs. Keep in mind that twelve hours of artificial light is equal roughly to four hours of natural light. Since plants need a range of wavelengths of light, use either a special "grow-light" or a pair of fluorescent bulbs, one labeled "warm white," the other, "cool white." Plants with flowers should be approximately 15 cm away from the fluorescent light source. Leafy plants can be as much as 30 cm away. African violets probably grow better under artificial light than any other plant. Another plant that thrives under artificial light is the begonia.

Reference books tell you the most favorable lighting conditions for your plants. Desert plants should get full sunlight from the south. Also, plants that produce flowers (such as geraniums) need full sunlight. Plants native to the floor of tropical forests tend to do well with light from windows that face east. Plants with large foliage leaves such as umbrella plants and some ferns do nicely with light from the north. And so on.

One clue that your plants are getting too little light will be yellowish leaves rather than green ones. Another symptom is long, skinny stems. Furthermore, the plant may lean excessively toward whatever meager light is available. If a plant gets too much light, the entire plant will take on a yellowish color and the leaves may appear "scalded" or burned. Since all plants bend toward the light because of the unequal growth caused by *auxins* (plant growth substances, often called "plant hormones") they produce, you will have to turn your plants around from time to time if you want them to grow straight.

IT'S NOT THE TEMPERATURE, IT'S THE HUMIDITY

Your indoor plants will do much better if the temperature is low and the humidity is high, rather than the reverse. This is true for most house plants, since most of them originate in tropical rain forests, where there is lots of water.

You can add moisture to the air by having trays or pans of water near your plants. Many people who heat their homes with steam place trays of water on their radiators to keep the air moist. Their plants will benefit from this practice as well. Other people spray a fine mist of water into the air near their plants. You may not realize it, but your plants lose a lot of water through their leaves by transpiration (see Fig. 3.5). Additional water is lost through the surface of the soil through evaporation. If the flower pot is made of porous clay, even more moisture is lost.

The humidity in most homes is usually far lower than that in a tropical forest. If you want your plants to do well, keep the humidity near them high! Hot-air heating, steam heat, and even air conditioning and fans dry out the air. If you can, grow your plants in the kitchen or bathroom because these rooms are more humid than other parts of the house.

Orchids and other tropical plants thrive at lower-than-normal temperatures (as much as 10° C lower), provided the humidity is right. Windows are dangerous places to keep plants in winter because the temperature next to them may be 5° C cooler than in the rest of the room. Placing plants on radiators or heating units of any

Figure 3.5. Transpiration. Water droplets have collected on the right side of the bell jar.

kind is even worse, because the intense heat and drying effect will be fatal to most plants. One trick is to place a thick wooden cover over the radiator and put your plants on top of that. Some heat passes from the cover to the bottom of the plant container, but this can only be helpful to the plant. Don't use asbestos, because it is dangerous to your health.

REPOTTING

A healthy plant will probably outgrow its pot or container eventually. Here is an easy way to tell if your plant is ready for repotting. Turn the pot upside down and, supporting the plant with one hand, bang the rim of the pot on a table. You need to repot if the plant and roots slip out easily. All you need is a pot approximately 2 to 3 cm larger than the old pot. If you use a much larger container, the plant will exert all its energy and time filling the pot with roots; the stems and the leaves will grow poorly. Put fresh soil in the new pot.

TO POT OR NOT

The seeds of many common garden vegetables and flowers can be started in your home just as if they were house plants. In this category are almost all flowers, tomatoes, squash, cabbage, and many other vegetables. However, there are seeds that should not be started in pots and then transplanted to the out-of-doors. The reason is that many, if not all of them, will not survive the stress of transplanting. Many of the common vegetables are in this group, for example, beans, peas, carrots, and corn. If you are in doubt, read the label on the seed package.

TRANSPLANTING

Most experts recommend transplanting seedlings out-of-doors when they have their second set of true leaves. (The first pair are food leaves or *cotyledons*, which act as a picnic lunch for the young seedling until the true leaves can carry on photosynthesis and feed the seedling.) I recommend that you transplant the seedlings when the *first pair* of true leaves appears. The reason is that the smaller the plant, the less the shock it will experience when the roots are removed from the pot and inserted in the garden soil. Water the transplant at least once a day for four days after transplanting. Soil and air temperature, pH of the soil, and soil moisture must be within normal limits for your seedlings.

PLANTING SEEDS OUT-OF-DOORS—
SOIL AND OTHER MATTERS

Starting seeds out-of-doors is slightly different from starting them in your house or apartment. In practically all cases, the first thing you must do is improve the quality of the soil so your plants will be healthy and hardy. Good soil must have spaces in it so that it can hold air and moisture. Nothing will grow in hard-packed soil. Think of a heavily traveled path through a vacant lot! Dig up your garden plot and be sure to break up the large clumps of soil.

Most soils contain either too much sand or too much clay. Sandy soil has too many spaces between the grains, and moisture is easily lost. Soil with a lot of clay is the opposite: it has too few spaces in it. Humus, which is mainly decomposed plant and animal material, helps hold in the air and water and will improve both kinds of soil. You can make your own humus by a process called *composting.* (See the instructions on page 25 under the dirt science section.)

As long as you are adding to the soil, *don't forget our old friends, vermiculite and peat moss.* Even with humus, which is rich in minerals, it is wise to add fertilizer from time to time. Your growing plants will use up some of the minerals, and every time it rains, some will be lost by passing out of the topsoil region where the roots are. Minerals can be provided by mixing equal parts of compost (a mixture of decayed leaves and other plant material), manure, and humus. Spread 1 kilogram (kg) of the mixture on every 3 square meters (sq m). Some authorities recommend adding dried blood, wood ashes, and leaf mold to provide all the minerals that plants require. If you are interested in this topic, refer to the Bibliography at the end of this book, which lists several references.

FERTILIZE YOUR PLANTS

Fertilizer has to be added to the soil from time to time because the plants use up the nutrients and minerals as they grow. Nitrogen, phosphorus, and potassium are the three major minerals. Nitrogen is important in the formation of chlorophyll and gives leaves a deep green color. Phosphorus is necessary for good stem and root growth, while potassium is used by plants to make flowers. If you examine the label on a fertilizer container, you will see three numbers, for example, 5-10-5 or 20-5-10. The numbers indicate the percentages of nitrogen, phosphorus, and potassium, in that order.

Elements needed in very tiny amounts are called "trace" elements. Too much of them is worse than too little. The trace elements include calcium, copper, boron, iron, and zinc. These are usually in fertilizers but are not expressed by a percentage because the amounts are so small.

On the average, plants that are actively growing require fertilizer about once a month. Most vegetables are in this category and are referred to as *heavy feeders.* Very few house plants are heavy feeders and they absorb nutrients slowly over a long period of time.

SOIL pH

You may use the right ingredients to enrich your soil and still find that plant growth is poor. The most likely reason is that the pH (see p. 27) of the soil is too low and the soil is too acidic. Most plants thrive when the soil is close to neutral (pH range between 6 and 7). Unless the pH is within the proper range, the roots can't absorb the minerals in the soil. Flowers in this group include begonia, marigold, geranium, and carnation; vegetables are lettuce, radish, carrot, and tomatoes.

Most plants are acid-loving, though a rare few favor an alkaline environment, among them coltsfoot. Azalea, lily of the valley, peanut, and potato are some acid-loving plants. The proper pH range for azaleas, for example, is 4.5 to 5.5.

Use hydrion paper or a chemical pH kit to determine the pH of the soil. Both contain indicator dyes that change color as the pH goes up or down. The color your soil produces is matched against the colors on a key provided with the kit or the hydrion paper. The simplest device for measuring the pH of the soil is the pH meter, which has a range of approximately pH 4 to pH 9 and a probe attached to the meter. A soil sample is moistened with water and the probe inserted. After one minute has elapsed, the meter is read. Garden shops and scientific supply companies carry pH indicators, meters, and paper.

Changing the pH of soil is not an exact science, and most plants have a range for pH balances that is favorable to growth. If the pH is too high, adding extra peat moss or sphagnum moss is a way of lowering the pH. To lower soil 1 pH, add 1.5 kg of peat to each square meter of garden soil. If the soil has a pH that is too low, lime can be mixed with your garden soil to raise its pH. Adding 300 grams (g) of ground limestone to a square meter of soil will raise its pH by 1.

WATERING NEEDS FOR OUTDOOR SEEDS AND PLANTS

After you plant the seeds in the garden to the depth recommended on the package, water them well. Check by digging down to the seed level to be sure that the seeds and surrounding soil are always moist. During hot weather you may have to water once or even twice a day. If the seeds dry out after beginning germination, they will die. Carrots take two to three weeks to germinate, and some gardeners get discouraged and stop watering. As a result, the carrots which may have sent down roots without sprouting above-ground die.

How much water and when to water seedlings and mature plants depend on the drainage qualities of the soil and other factors. Soil with a lot of clay drains and dries out slowly. It doesn't have to be watered as much as sandy soil. Soil rich in sand must be watered frequently. Lettuce, carrots, and other plants with a shallow root system need a constant supply of water if they are to grow well. Tomatoes and corn, which have a deeper root system, do better with fewer but heavy waterings. Plant age is another factor. The younger the seedling, the more constant the water supply must be. Recently transplanted seedlings need a constant water supply since there is the shock of transplanting to overcome. The amount of sun and the wind blowing over your garden site must be taken into account when considering the watering needs of your plants.

STAY WARM—DON'T LOSE YOUR COOL

Seeds and mature plants need an optimal air and soil temperature in order to thrive or grow at all. Corn does well only in hot weather. Carrots, lettuce, and radishes can survive even in cold weather. Spinach does well in cool weather and grows poorly in the hot summer months.

In order for germination to take place, most seeds need a temperature around 23° C. You have two choices with seeds that must be started out-of-doors: You can wait for the climate and weather to cooperate with you and an average temperature of 23° C; or you can improvise ways to beat the weather. Naturally, if you start the germination process in the summer, late spring, or even early autumn, you will probably have no problem with temperature.

During the winter and in cold climates, however, you will need to find a way to meet the temperature requirement of your seeds. If your home is warm all the time, the easiest thing is to place the seed containers where the temperature is usually close to 23° C. A closet or cabinet on an inside wall is a good place. Avoid window ledges, because they are often drafty and cold, particularly at night.

If you live in a private house and have access to the boiler room, try the following method (this is not meant for apartment house boilers but for one- or two-family homes). The top of the hot-water heater will work out fine as your incubator. Figure 3.6 shows what the setup looks like. Place a few ceramic dishes or building bricks on top of the boiler. Use these to support your germination containers and watering tray. The supporting structures form an air space that will heat the water in the tray without cooking the seeds. Don't place the seeds over steam or you will kill them.

HOT AND COLD FRAMES

A *cold frame* (see Fig. 3.7) is an enclosed bed of soil used to start seeds outdoors, and it is heated by the sun. The cold frame is merely a bottomless box that extends about 25 cm into the ground. A cold frame will let you get an early start germinating your seeds out-of-doors. You can buy or make a cold frame from scrap wood and use either transparent plastic or glass for a cover. The wood

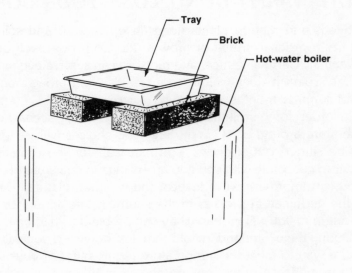

Figure 3.6. Homemade seed hot plate

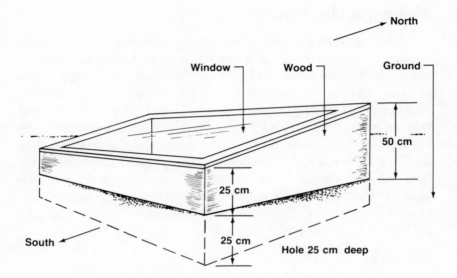

Figure 3.7. Cold frame made from an old window and scrap wood. The slanted top allows precipitation to run off easily.

serves as the framework, and the transparent material on top allows the sun to enter and heat the soil inside. What you have is a small greenhouse.

Making a Cold Frame

The size of your cold box can vary greatly and may depend on the material you use. The cover is slanted so that the highest edge faces north; the north end should be approximately 50 cm above ground level, the south end only 30 cm above the ground. If you can, use cedar or redwood for the frame because these woods resist moisture. Dig a 25-cm hole where your cold frame will be. Replace that soil with your own mix. Your cold frame warms the soil only when the sun is shining. When the sun is strong and the air temperature is warm in late winter, check the inside of the cover. Look for condensed moisture on the inside of the transparent cover. If there is some, open the cover until the evening. If you don't do this, your seedlings may die.

Making a Hot Frame

You can turn the cold frame into a *hot frame*. The difference between a cold frame and a hot frame is that the cold frame is heated by the sun, while the hot frame has a supplementary source of heat, often electric wires buried in the ground. Most seed companies sell soil-heating cables which you can bury in the soil inside the cold frame in a suitable sunny area. The cables have a thermostat so that the seeds are not roasted. The cables also are waterproofed, so there is no danger of getting a shock if you follow directions. Bury the cables about 5 cm beneath the soil. Now, fill the box with your soil mixture, add the seed, water, and close the cover. Finally, turn on the electricity. Now you are in the seed germination business, independent of the season of the year.

4   PLANT GROWTH EXPERIMENTS

GETTING STARTED

You may have noticed that the last chapter did not offer any suggestions for experiments or plant projects. The reason is that you can't embark on a plant experiment or project unless you can germinate seeds and care for your plants. They must be healthy and survive if your experiment is to have any meaning. Now that you know how to do this, you are ready for projects and experiments. The experiments in this chapter deal with insects, seeds, air, soil, photosynthesis, pollutants, and more!

Do Marigolds Really Repel Pests?

Marigold seeds are often grown along with vegetable seeds in gardens because marigolds are believed to keep pests away. In particular, they are supposed to repel nematodes (eelworms), microscopic worms that live in the soil and feed on the roots of plants. The plants they attack—tomato roots are a favorite—turn yellow and grow poorly. Marigold and tomato seeds could thus be the basis of an experiment on pest control.

Plant tomato seeds in a section of your garden and surround them with a border of marigolds grown from seed. As a control, grow the same number of tomato plants in a similar section of your garden. The control group, you may remember, must be as similar as possible to the experimental group except for one factor: In this case, the experimental factor is a border of marigold plants. Make sure no marigolds are growing near the control tomato plants. Do marigolds keep nematodes away from tomatoes? Do marigolds keep other pests away?

You can try one or more of the following projects or try to devise an experiment to answer the questions.

● Do insect pests avoid living near marigolds? Set up two groups of the same vegetable (tomatoes) and sprinkle ground-up heads of marigold flowers around one group daily. How does the amount of insect damage done to this group compare with that done to a control group of vegetables?

● Instead of ground-up flower heads, try, in turn, ground-up marigold leaves, stems, and roots. Which part is most effective?

● Do insects avoid the *odor* of the marigold flower or a special *chemical* produced by the roots of the plant and excreted into the soil?

● What pests, other than nematodes, do marigolds protect tomatoes from?

Reportedly marigolds reduce the number of cabbage worms that feed on cabbage heads. Set up a similar experiment to the one with the tomatoes, using cabbages instead.

FLOWERPOT EXPERIMENTS

Is seed growth affected by a pot's shape and what it is made of (see Fig. 4.1)?

Try the following experiments or others of your own devising:

● Grow African violets in square pots and in round pots that have the same area. How does the appearance of the leaves and flowers differ? Can you discover what causes these differences? Repeat this experiment using other plants that are easy to grow in pots, for example, cacti, philodendron, or the prayer plant.

● Do cacti grow better in a wide or narrow pot? Is there a ratio between pot size and stem diameter that determines the width of an ideal container?

Figure 4.1. African violet. Would this plant look differently if it were growing in a square pot?

● Will a particular plant grow better in a pot made of plastic or one of clay?

● Is a clear plastic pot better than an opaque plastic container to grow plants in?

SEED EXPERIMENTS

Some seeds have a hard coat that prevents water from entering easily. Seeds of this type have to be *scarified*—have some of the waterproofing material in the seed coat removed. Do this by carefully scraping the large end of the seed with a razor or a file; you want only to remove *some* of the coating. If you remove too much and penetrate the seed, you may damage the embryo so that it will not germinate. Seeds that must be scarified include geranium and morning-glory.

Here are some experiments to try and questions to answer.

• Using either of the seeds just mentioned, compare the germination time necessary for scarified seeds and untreated seeds.

• Can a more efficient tool be used? Is a nail clipper better than a file?

• How does scalding the seeds (pouring boiling water over them) affect germination time?

• Soaking the seeds is another alternative to nicking the seed coat. The soaking time will vary from an hour or so for geranium seeds to several weeks for seeds of the Kentucky coffee tree. Which method is best?

• What is the effect of chemicals on seed germination? For starters, try germinating a batch of seeds after soaking them for a week in one of these: bleach water (Clorox™) (when using bleach, be sure to wear goggles and wash hands thoroughly), hydrogen peroxide (3%), vinegar, sodium bicarbonate (5%), citrus fruit juices (tomato, orange).

• Young seedlings manufacture their own vitamins and other organic compounds by chemically uniting minerals in the soil to the products of photosynthesis. Add vitamins and proteins in soluble form to the starting mix. Set up a control group. Is there any advantage to adding vitamins A, C, D, etc., to the starting mix? What happens when you vary the dosage? What happens to seedling growth if combinations of vitamins are used?

• How should seeds be stored to remain alive for long periods of time (five years or more)? Try varying amounts of cold temperatures or different kinds of airtight or lightproof containers.

SEED PROJECTS

The number of seed projects you can do is unlimited. As you work with your seeds, new ideas will surely pop into your head. Here are a few ideas for projects:

• Discover a way to "harden off" plants indoors. "Hardening off" refers to the process of taking seedlings started indoors and leaving them outside for increasing periods of time over several days so that they get used to their new environment before being transplanted.

• Find a new and cheaper substitute for vermiculite or peat moss (plastic packing material and coffee grounds?).

• Develop a special light filter that will speed up chlorophyll development in newly sprouted white shoots. Try purple, red, and blue.

• Will seedlings grow better if exposed at night to ultraviolet light (from a special bulb used to illuminate posters)?

• Develop or invent a starting mixture that is better for germination and growing seedlings than commercial preparations.

• Scientifically compare several commercial seed-starting mixes. Which is the best?

• Discover or develop a new source of heat for the hot frame. Some experiments have been done using fermenting corn stalks or manure instead of electricity.

• Can solar heat be used in connection with your miniature greenhouse?

• Find new ways for growing difficult plants at home. We already spoke about celery and peony plants. Another tough one to grow from seed is the rose. Many rose seeds are sterile. Fertile rose seeds have to be kept in moist peat moss for at least three months at 5° C. Orchids are another difficult plant to maintain at home. Starting them from seeds is even more complicated and is often unsuccessful.

SOIL AND WATER EXPERIMENTS

• Can you formulate a better special fertilizer and soil mix for plants with special needs, such as gardenias, which need an acid-type fertilizer?

• Is bone meal (an organic phosphate product favored by organic farmers) superior to superphosphate as a source of phosphorus for plants?

● Which is best for starting young seedlings—peat moss, humus, compost, or leaf mold?

● Can you prove scientifically whether it is better to water house plants by adding water from the top or by adding the water to the pan that the plant and its container sit in?

● Will plants grow larger and be healthier if given tea or coffee instead of plain water?

● Can you devise an automatic watering system that will provide a constant supply of water so that a plant will be neither waterlogged nor too dry? The two most common systems use wicks that conduct water from a reservoir or pots in a tray filled with water.

PHOTOSYNTHESIS EXPERIMENTS AND PROJECTS

Our world runs on energy from the sun. Just think of oil and gasoline! They are the remains of plants and animals that lived long ago. Ancient animals fed directly on green plants or on animals that ate green plants. The story is exactly the same today. Green plants use their chlorophyll to change inorganic carbon dioxide and water molecules into organic molecules for growth and energy. All animals, including humans, either feed directly on green plants or feed on animals that ate green plants. As the green plants make their nutrients, they release oxygen into the atmosphere. Briefly, this is a description of *photosynthesis*.

Scientists are keenly interested in fully understanding photosynthesis because all of our foods result from photosynthesis, either directly or indirectly. Can you think of a food that can't be traced back to a green plant?

Besides furnishing us with food, photosynthesis replaces the carbon dioxide in the atmosphere with oxygen (see Fig. 4.2). Without a constant source of oxygen, almost all living things on earth would die, including you and me! Research on photosynthesis is going on all over the world.

Right now, only intact, living plant cells can carry out all the steps in photosynthesis. If we could carry out this process using just

Figure 4.2. Light provided energy so that this tree could grow to this size.

chlorophyll molecules in laboratory glassware, food could be made anywhere there was light.

Most photosynthesis research is being done by biochemists at the molecular level. These experiments require the use of expensive equipment and radioactive tracers. Experiments on this level are beyond the means of the average home or school researcher. However, you can do some photosynthetic experiments that call for relatively simple equipment. One such experiment follows.

Determine What Carbohydrates Are Produced by Plants

For this experiment you will need:

healthy green plant (coleus)
black paper
aluminum foil
knife

dilute iodine solution (prepare by pouring a small amount of iodine, which most likely is in your medicine cabinet, into a small container, and add water until the solution is a pale yellow color)

1. Select two leaves as similar as possible. Wrap one leaf with black paper.

2. Cover the black paper with aluminum foil to insure that no light reaches the leaf. This is your experimental leaf.

3. Do nothing to the other leaf. That is the control.

4. Let the plant remain in its normal place for several days. The covered leaf will not be able to manufacture food and will use up its stored food.

5. Remove both the experimental and control leaves.

6. Keep each in a separate marked container so you know one from the other.

7. Boil the leaves in water for five minutes.

8. Remove the chlorophyll. This is done by boiling in alcohol. Since alcohol ignites easily, boiling in a water bath is the safest method.
Fill a large pot or a 500-ml beaker one-quarter full with water. Next, pour alcohol into a small pot or 250-ml beaker half-filled. If possible, use 95% alcohol since this will dissolve the chlorophyll best. Place the alcohol container in the larger one. A hot plate is the preferred method for heating. If a hot plate is unavailable, you will have to heat the water bath over an open flame. *Exercise caution and have an experienced person assist you because alcohol boils at a low temperature (lower than water) and ignites easily.* Boil each leaf in the alcohol for five minutes.

9. After boiling, remove the leaves. Place them in separate, labelled dishes.

10. Add the iodine solution to each leaf. Iodine will turn starch black. Which leaf contains starch? Which regions of the leaf store starch?

Here are two spin-off experiments:

● Using Clinitest™ paper or Benedict's solution (available in drugstores and from chemical supply companies), test both leaves for simple sugar. Follow the directions on the Clinitest™ paper container. Chop up a leaf or grind it in a mortar. Place the leaf pieces in a test tube filled one-quarter of the way with Benedict's solution. Heat the tube and its contents for four or five minutes. The solution should boil gently. If the solution changes to a brick-red color (starting color is blue), sugar is present.

● If you have access to the appropriate chemicals and equipment, test the leaves for specific sugars (dextrose, mannose, etc.).

Measure Photosynthesis Rate by Measuring Oxygen Production

The amount of oxygen a plant releases is one way you can measure its rate of photosynthesis. This is easily done using a freshwater plant such as elodea and the setup shown in Figure 4.3. The number of bubbles given off in a set period such as a minute can be calculated using a watch or clock.

1. Use a healthy piece of elodea. Place it upside down in the test tube.

2. Fill the test tube with a 0.2% solution of sodium bicarbonate (baking soda). To prepare this solution, dissolve 1 g of sodium bicarbonate in 500 ml of water.

3. Place a piece of rubber tubing 7 cm long over a 10-cm length of glass tubing inserted into a one-hole stopper.

4. Fill the glass tubing with the sodium bicarbonate solution. Do this by placing the rubber tubing in your mouth and sucking up the solution. Quickly clamp the rubber tubing with a pinch clamp in order to keep the solution in the tubing.

5. Carefully place the lower end of the glass tubing over the end of the elodea stem.

6. Push the stopper in place to seal the setup.

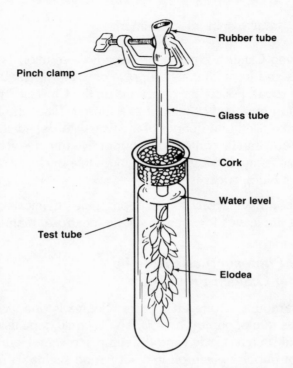

Figure 4.3. Setup for counting oxygen bubbles liberated from elodea

7. Place the setup in a dark room so that you can control the light.

● What happens to the rate of bubbles released per minute when you use a 100-watt light positioned 25 cm from the elodea plant compared with a 40-watt bulb at the same distance?

● What happens to the rate of bubbles released per minute if you use a 100-watt bulb and change the distance to 50 cm? 100 cm?

● What is the effect upon the bubble rate if you use a 60-watt yellow bulb? A 60-watt blue bulb? A 60-watt "poster" (ultraviolet) bulb?

● Can you devise a mirror arrangement that reflects the light from the 100-watt bulb back through the elodea leaves again to increase the amount of photosynthetic activity? Most of the light that strikes leaves passes through them. Only a small percentage of the light energy is involved in the process of photosynthesis.

• Can you increase the rate of oxygen bubbles that are given off without increasing the intensity of the light?

• Use the same experimental setup of mirrors in conjunction with blue or red cellophane or glass. Also try green cellophane or glass. Which color(s) is/are best for photosynthesis? Which is/are the worst?

Here are some related ideas for projects:

• Will vegetables and flowers (leaf lettuce, tomatoes, zinnias, cosmos) die if they receive too much sunlight (more than 18 hours per day)? For this experiment, you will have to expose the experimental plants to light after sunset. By using timers, you could first try 18 hours for two weeks and note the effect upon the flowers or vegetables. Then try 20 hours, next 22, and finally 24 hours.

• Is there a critical light period for peak flower or fruit production?

• Is there a critical light period that causes the plants to die?

• Will red light (use red bulbs or colored photographic filters and plants growing in a darkened room) produce long, spindly plants that can't support themselves?

• Will blue light produce small, short plants (marigolds, zinnias) that fail to produce flowers?

IS THE LIGHT TIME THE RIGHT TIME?

I am allergic to ragweed pollen, so I take my summer vacation in Maine whenever I can: Ragweed doesn't grow in northern Maine. A ragweed plant produces no flowers until the daylight hours are less than fourteen and a half. In northern Maine, this happens only near the end of August. Since Maine's climate is cold, the first frost usually comes early in September, if not before. Since frost kills ragweed flowers, in parts of Maine ragweed flowers have no time to produce seeds.

This dependence on a particular period of daylight before producing flowers is called *photoperiodism*. Most plants will not flower until they have reached a certain size or age. Once a plant is large enough or old enough to flower, its flowering often depends on the length of the day. Flowering plants can be separated into three groups: day-neutral, long-day, and short-day. Day-neutral plants produce flowers regardless of the number of hours of light. Peas, garden beans, roses, and tomatoes are examples of day-neutral plants. Long-day plants are those that start to produce flowers when the number of daylight hours *exceeds* a certain minimum value. Long-day plants bloom in the spring and early summer. Beets, radishes, spinach, and clover are a few of the long-day plants. Spinach won't flower unless the light hours exceed fourteen and a half. You couldn't grow spinach outdoors near the equator and expect to get flowers because the days are around twelve hours long there. Short-day plants, on the other hand, start flower production when the length of the day becomes *less* than a specific length. These plants bloom in the early spring and fall. Short-day plants include aster, cocklebur, goldenrod, strawberries, and ragweed.

Oddly, whether a plant is a short- or long-day plant actually depends on *darkness* rather than on daylight. Uninterrupted darkness is the key. Even a brief flash of light during the dark period of only a quarter hour more or less than the critical amount will prevent a particular species of plant from flowering. Although we should speak of *long-night* and *short-night* plants, the early terms referring to *day* are still used. Since scientists are concerned with flowers and fruits, they have performed some interesting experiments that involved grafting day-neutral, long-day, and short-day plants to each other. These experiments led to the discovery of *phytochrome*.

Flash-dance

Phytochrome is a pigment sensitive to red (wavelength = 660 nanometers [nm]; a nanometer = one billionth of a meter) and far-red light (wavelength = 730 nm). Phytochrome is activated by red light and deactivated by far-red light, which is invisible to our eyes. Phytochrome "tells" the plant if it is in light or in darkness, but some sort of internal clock measures the time between darkness and the next exposure to light.

Biologists have discovered that a flash of red light destroys the effect of the dark period and that a flash of far-red light destroys the effect of the red light. If the dark period of a short-day plant, which normally would allow the plant to flower, is broken by a flash of red light, the plant will not flower. However, if the flash of red light is followed by a flash of far-red light, the plant *will* flower. You could flash red, then far-red again and again, but whether or not the plant flowers depends on the last flash.

Is Florigen for Real?

Experiments have led some scientists to believe that after the "starting" signal in terms of the period of darkness, something else must be responsible for the development of a flower. In some cases only one leaf need be exposed to the proper light period in order for the plant to make flowers. This suggests that some chemical "messenger" is sent from the exposed leaf to all the stems, "telling" them to begin flower production.

The cocklebur is a short-day plant that has been widely used for experiments with flowering. Plant scientists have covered a *single* cocklebur leaf on both sides with a light-proof covering for nine hours. This is the critical time period for flowering for this plant. The rest of the plant was exposed to only eight and a half hours of darkness.

Surprisingly, flowers will develop on *all* parts of that cocklebur plant. More remarkable, if we graft that one leaf exposed to the nine-hour critical darkness period onto a different cocklebur plant— one not exposed to nine full hours of uninterrupted darkness—that plant, too, will flower! Graft some leaves from that second cocklebur onto a third and a fourth cocklebur that had too little darkness and they too will flower, just as if they received the full nine hours! Figure 4.4 shows how this experiment is done.

From this experiment came the belief that a hormone was responsible. So far, however, no one has been able to find it. Other

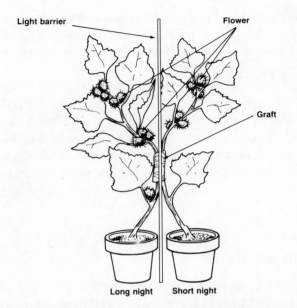

Light barrier   Flower

Graft

Long night   Short night

Figure 4.4. Grafted cockleburs. If two cocklebur plants are joined by grafting, exposing only one of the plants to the proper light period will induce flowering in both plants.

plant scientists believe the hormone, named florigen, is not one substance, but two or more known plant growth substances such as IAA, gibberellic acid, cytokinin. (See chapters 5 and 9 for growth hormone experiments.) Still other scientists believe that florigen is some kind of virus or is a virus-like particle. Can you solve the mystery?

Most likely, you don't have the equipment to try to find out if florigen is a virus. You *could* design an experiment to determine whether florigen is a combination of known, available growth substances (see chapters 5 and 9). Here is one possible experimental hypothesis: "Will flowering occur in cockleburs if they are given *cyocel* (an available plant substance that blocks the formation of gibberellin)?" If flowers do not form, this would indicate that gibberellin is probably one of the substances that compose florigen. This experiment is probably too complex and demanding to do as your first plant experiment. If you do try it, have an experienced researcher review your experimental plan.

Light Projects

• Determine if a certain flowering plant is a long-, short- or neutral-day plant.

• Find out if there is a common feature for short-day or long-day plants other than the time of the year they flower.

• Some seeds need to be exposed to a particular color of light before they will germinate—lettuce, for example. Try to discover what color light improves the germination percentage, or shortens the germination time of lettuce. Use colored cellophane, glass, or plastic to screen out unwanted colors.

BIOLOGICAL TIME

Biological clocks in plants are a fascinating field for experimentation, even though we still do not know exactly what a biological clock is. The study of biological clocks is not new. They have been studied for more than 50 years. We can define a biological clock as an internal mechanism for measuring time by organisms using the rhythms of their life processes.

How accurate are biological clocks? We know that some are accurate to within 15 minutes in telling the length of darkness. Biological clocks induce some plants to release their nectar only at a certain time of day; others induce flowers to close their flowers at a certain time of day. The four-o'clock gets its name from this type of behavior. Can you get four-o'clocks to stay open later by using auxins or other chemicals? Try this project only if you feel confident in your scientific abilities.

TRANSPIRATION AND HUMIDITY EXPERIMENTS

Earlier we discussed *transpiration*, the process by which plants lose water through their leaves. We also discussed how plants lose water through porous clay pots and from the surface of the soil. In the following experiments, you can investigate water loss from plants.

Water Loss: Amount

Determine the amount of water lost by a plant due to transpiration.

1. Use a well-watered potted plant. Drape a large piece of clear plastic over the leaves. Pick up the edges of the plastic and enclose the leaves with the plastic.

2. Tie the plastic around the base of the stem. Make an airtight, watertight seal by wrapping the plastic around the stem several times. Seal the plastic with transparent tape.

3. After several days, moisture will have collected inside the plastic.

4. To determine how much water was lost from the leaves, do the following:
Weigh a small, dry beaker.
Remove the plastic and collect the water by running it off the plastic into the beaker.
Weigh the beaker with the water in it. How much water did the plant transpire?

Water Loss: Location

How much water is lost through the sides of a pot and from the soil?

1. Completely dry a clay pot and its soil in the oven by heating at 200° F for one hour.

2. Add a weighed amount of water to the soil—enough to make it soggy.

3. Allow the pot and its soil to remain exposed to the air in a room of your house for five days.

4. At the end of five days, weigh the pot and the soil. Record the weight. How much water was lost in those five days?

5. Repeat the experiment using a nonporous plastic pot. Heat only the soil, not the plastic pot!
Compare these results with those for the clay pot.

Comparative Transpiration

Compare the transpiration rate of a cactus with that of a small plant like *Sinningia pusilla* that will fit easily inside of a coffee cup.

1. Start with a cactus and *Sinningia* plant, which have the same stem diameter and height.

2. Wrap the leaves and part of the stems that are above ground with plastic. Do this to both plants. Expose to indirect sunlight for four days.

3. Collect the water from each plastic wrap in separate beakers whose weight was previously determined. Compare the amount of water lost in each case. Which plant lost more water? What happens when the plants are placed in *strong* sunlight for four days? Why do you think cacti thrive in deserts?

Relative Humidity

Relative humidity is defined as the amount of moisture in the air compared to the amount of moisture the air can hold at a certain temperature. Relative humidity is expressed as a percent. When the relative humidity is high, 80% or more, the air will not be able to hold much more moisture. Temperature is an important factor in determining relative humidity. The warmer the air, the more moisture it can hold.

To determine the ideal relative humidity for a particular plant species such as coleus, you will need one piece of special equipment: a wet-and-dry-bulb thermometer. You will also need a relative humidity table, which you can find in a book on meteorology (weather). (Look up *relative humidity* in its index.)

1. Grow members of one plant species in identical environmental conditions of light, temperature, and care. Keep everything the same except for the relative humidity.

2. Vary the relative humidity by spraying the plants once or more each day with a plant mister. A clean glass-cleaner bottle works fine. The more times you spray, the higher the relative humidity.

3. What is the effect on leaf production and size if you spray

one plant once a day for four weeks? Another plant twice a day for the same time period? Another plant three times a day? Don't forget that you will need a control that isn't sprayed at all. In each case, determine the average relative humidity for the two week period using the wet-and-dry-bulb thermometer and the relative humidity table.

4. Observe and record the number of new leaves formed on each plant during the two-week period. Measure the size of the leaves of each plant. Start your measurements at the tip of the leaf. End where the *petiole* (the stalk that connects the leaf to the main stem) begins.

5. Rooms that normally have higher relative humidities than other parts of your home are the kitchen and bathrooms. If you have a sunny bathroom, try growing the lipstick plant (*Aeschynanthus lobbianus*) and pandanus (screwpine) there. Which grows better?

6. Experiment with other plants. Try growing them in sunlit bathrooms or kitchens. Prepare a list of common household plants that grow well in sunny bathrooms and kitchens.

PLANTS AND AIR POLLUTANTS

As our civilization becomes more technical and industrialized, our atmosphere becomes filled with more polluting gases. Factories, power plants, cars, and planes are some of the sources of these pollutants. Even statues and grave markers are being eroded by the exhaust gases, particularly the oxides of nitrogen and sulfur. If these gases can destroy stone, what do they do to plants? You will have a chance to find out. Since these gases are poisonous, caution will have to be exercised when carrying out some of the experiments and projects in this section. The first project is a relatively safe one.

Survey the plants growing near a factory that produces air pollutants such as nitrogen oxides (any burning causes nitrogen in the air to combine with oxygen in the air) and sulfur oxides (from sulfur-containing fuels).

Compare common (e.g., dandelions or clover) plants growing near the factory with members of the same species growing in a nearby nonpolluted area. Or, compare plants growing along the side of a road with those growing in a neighboring place where they are not exposed to the exhaust fumes of autos and trucks.

Compare the size of the plants in each area. Which group has more flowers? Is there a zone where no plants grow? If so, can you draw a map to scale indicating the factory, the no-growth zone, limited growth zone, etc.?

Sulfur Dioxide Experiments

Working with pollutants and acids in the following experiments can be hazardous. *Do these experiments only under the **direct** supervision of a scientist or science teacher. Otherwise, avoid them!*

Other pollutant gases that enter our atmosphere are ozone and carbon monoxide. Avoid working with these two because they are extremely dangerous! I believe that no young researcher should attempt any experiments with them even if supervised. There are other less dangerous pollutants you can work with.

Assuming you have proper supervision, first build or buy a gas-tight growth chamber in which to expose your plants to the pollutants. The chamber should be of clear plastic so that the plants can receive sunlight. Obviously, it must be airtight so that the pollutants won't escape and be inhaled by you! Only your experimental plants should be exposed to these pollutants. You also need a small pump— e.g., the kind used in fish tanks—and a flow meter to measure and regulate the flow of gases. How this equipment is to be arranged is shown in Figures 4.5 and 4.6.

Sulfur dioxide (SO_2) is a colorless gas contained in the smoke produced when any sulfur-containing fuel such as coal or natural gas is burned. Sulfur dioxide forms sulfuric acid (H_2SO_4) by combining with the water in moist air.

To make your own sulfur dioxide, add a few drops of concentrated sulfuric acid to 5 g of sodium sulfite dissolved in 50 ml of demineralized water placed in a gas generator setup (see Fig. 4.5). *This should be done using a chemical hood with proper venting and a fan to carry away noxious fumes.*

For your experiment to have meaning, you must set up control plants in identical airtight growth chambers. Pump sulfur dioxide

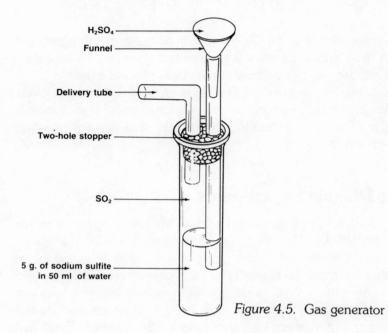

Figure 4.5. Gas generator

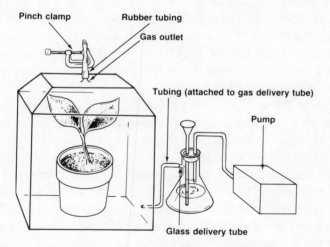

Figure 4.6. Pollution chamber attached to gas generator by rubber tubing

into your gas-tight plant growth chamber at the rate of 1.5 liters per minute for three minutes daily.

Observe your plants over five to seven days in order to collect enough data to draw meaningful conclusions. You can compare the control and experimental plants in terms of number of new leaves, size of flowers, appearance of leaves and stems, number of leaves that fall off. Of course, you should develop your own criteria in place of or in addition to these.

Table 4.1 can be used to select plants that you may wish to use in your sulfur dioxide experiment. The number of plants in each category is rather brief. Through your experimentation, can you add more plants to this table?

Table 4.1

SULFUR DIOXIDE SENSITIVITY

Sensitive Plants— Can't tolerate 0.3 ppm (parts per million) of sulfur dioxide	Average Plants— Can tolerate up to 0.3 ppm of sulfur dioxide	Resistant Plants— Can tolerate more than 0.3 ppm of sulfur dioxide
Squash	Iris	Rose
Oats	Aster	Onion
Lettuce	Marigold	Corn
Carrot	Cabbage	Lilac
Turnip	Begonia	Virgina Creeper
Salvia	Parsley	
Radish	Gladiolus	
Cosmos	Tomato	

Nitrogen Dioxide Experiments

Nitrogen dioxide (NO_2) has a brownish color and is one of the main ingredients of car exhaust. It can be made by placing several copper strips in the gas generator described in the previous section and adding concentrated nitric acid (HNO_3). Follow a similar procedure for that suggested for sulfur dioxide (pump in 1.5 liters per minute of nitrogen dioxide for a three-minute period daily). Some plants are resistant to nitrogen dioxide (cucumber, cabbage, and gladiolus); others are quite sensitive (spinach, celery, romaine lettuce, and petunia). Experiment with one or more of these plants and see if your results agree.

What plants can you add to the list of sensitives? Of resistants?

Pests and Pollution

Research indicates that air pollution may be bad for plants but wonderful for the pests that feed on them. Aphids grow faster over a three-day period on plants exposed to both sulfur dioxide and nitrogen dioxide than on plants exposed to nonpolluted air. You can investigate this by means of the kind of relatively simple experiments described here.

1. Grow six or more similar tomato seedlings in identical airtight growth chambers under the same environmental conditions. Infect plants in both chambers with 25 aphids each.

2. Measure the average size of the aphids. Keep accurate records.

3 (a). Prepare both sulfur dioxide and nitrogen dioxide, following the directions given in the previous section. Using the pump, introduce both gases at the rate of 1.5 liters per minute into one chamber for three minutes daily for five days.

(b). Pump atmospheric air into the second chamber for three minutes daily for five days.

4. Record the average size and number of aphids on each plant daily.

Do pollutants cause insects to grow larger? Live longer, healthier lives?

Try this experiment with other plants and with different insect pests such as mealybugs or cabbage moths.

5   BITS AND PIECES

Let's suppose I have an unusual begonia plant with unique leaves. You say this begonia is just what you need for your project on photosynthesis. Not only that, you need at least forty such plants for your experimental and control groups. Your request can be met, and in addition, I can keep my original plant!

Plants and simple animals have a wonderful survival mechanism that higher animals lack. Under the right conditions, a piece of a plant, say, a leaf or a stem removed from a healthy plant, will grow its missing parts. Imagine an amputated finger growing a completely new body consisting of a heart, lungs, legs, brain, and so on!

Many plants can be produced by using small bits and pieces from one plant, and all will have the same genetic characteristics as the original plant. The reason is that there is only one parent rather than two as in the case of plants that reproduce by seeds. This lack of inherited variation offers some wonderful experimental possibilities to both the plant breeder and the researcher. The researcher is sure that the genes for all the plants in the experimental and control groups are identical.

People and higher animals resemble their parents because each parent contributes half the genetic material that determines what the offspring will be like. For example, you may have hair similar to your mother's and people may say that your eyes are exactly like your father's.

When two parents are involved in the reproductive process, the process is called *sexual reproduction* and the offspring will have some traits of each parent. Sexual reproduction involves sperm cells from the father and egg cells from the mother. If there is only one parent, the process is called *asexual reproduction*. This is the same as a new individual developing from only an egg cell or only from a sperm cell. The offspring will have the identical genetic material as its sole parent cell. If a plant has some special trait and a piece of that plant grows into a completely new plant, the new plant will have genetic material only from a single parent, including that special trait!

Since seed formation involves *two* parents (sperm and egg cells), there is always the chance that a special trait may not appear in the offspring. It may be hidden by a dominant gene contributed by the other parent. Another possibility is that some undesirable trait could be introduced into the offspring by the other parent. For example, two corn plants may be crossed, both with genes for large, sweet kernels. One may have genes for poor root formation and the other may have genes for good root formation. The offspring may inherit poor root formation along with the excellent kernels.

The only way to grow *seedless* plants, of course, is by using a part of the plant (root, stem, etc.) that produces seedless fruit. You can't plant the seeds of a seedless grape vine. There are none!

Also, growing plants from seeds is usually a slow process, especially when the seeds are small. A good example is the strawberry plant, whose seeds are inside the small dots you see on the outside of the strawberry. Germination time for strawberry seeds is 30 days, while marigold seeds, which are much larger, germinate in only five days. After germination a minimum of four to six additional weeks must pass before you have a mature plant. The potato also produces tiny seeds. It is almost always faster and surer to start plants using the parts of an adult plant. Just think back to the last chapter on seeds and recall some of the reasons why seeds may fail to grow.

VEGETATIVE PROPAGATION

Using the roots, stems, or leaves of a plant to produce whole new plants is called *vegetative propagation*. Some plants do this on their own. Strawberry plants and many grasses spread throughout a garden plot by producing *runners*. There may be a spider plant (Fig. 5.1) in a hanging pot in your home with several *baby* plants dangling over the edge of the flower pot. The babies are attached

Figure 5.1. Spider plant. "Spider" is in the name of this plant because of the spiderlike appearance resulting from the runners that form the "legs."

to the parent plant by a thin stem (runner). Place any of the babies on soil. Give them water and light, and the small plants will take root and grow into large plants.

Other plants normally produce *rhizomes,* which are similar to runners except that they travel beneath the soil. If you have a healthy snake plant growing in a flower pot that has some room in it, soon you will see a new plant break through the soil near the original plant (see Fig. 5.2). After a while, another new plant will emerge from another rhizome until the container is filled (Fig. 5.3).

Bulbs, tubers, and corms can also be used to grow new plants. (Tubers and corms are very similar to bulbs.) The onion is a familiar bulb. Bulbs are underground storage and reproductive structures of certain plants. The central part of a bulb forms the flowers (reproductive organ) while food is stored in the fleshy leaves that surround the flowering shoot. Bulbs are planted in the autumn for spring flowering. After flowering, the plant makes and stores food in the bulb during the warm months so that it can flower again next spring.

Figure 5.2. Snake plant removed from the soil to show its rhizome

Bulbs can also be *forced* into flowering during the winter or at other times of the year, in other words, fooled into acting as if winter had just ended and spring arrived. You can then do a winter project with flowers if you have limited time. The easiest bulb to force is the paper white narcissus. Here are the directions for forcing this bulb:

1. Put the bulbs in a container, roots down. Hold the bulbs upright by adding a mixture of potting soil, vermiculite or clean gravel. Use equal proportions of each, and cover the roots by no more than 1 cm.

2. Add water to fill the container up to the base of the roots.

3. Place the container in a cool dark place (about 15° C).

4. Check every few days. When stem growth begins, move the bulbs into the light.

5. Watch for the flowers to appear in six to eight weeks.

Figure 5.3. A pot filled with snake plants

Some bulbs, such as the crocus, require six weeks of refrigeration before flowering. Many gardening establishments sell precooled bulbs (expensive) that are ready for forcing.

Projects with Vegetative Propagation

• Can chemical or hormone (auxin) treatment replace cooling for crocus bulbs?

• Will adding to the water a liquid fertilizer containing 10% nitrogen, 6% phosphorus, and 6% potassium cause bulbs to flower twice in one season?

• Are plants grown from seeds of forced flowers different from those that grow from spring flower seeds? How are they different? Why?

• Can *you* produce any new varieties by cross-pollinating (see chapter 7) flowers of forced bulbs?

● Plant a clove of garlic and see what happens. Start with one from your grocery store or local vegetable market. Garlic that you eat comes from a garlic bulblet or clove that surrounds a parent bulb. If you separate each bulblet and plant them, each will produce its own set of bulblets. Garlic bulblets are useful for cooking. Separate the cloves and plant them about 3 cm deep with the root end down. Usually garlic grows best if the cloves are planted in the spring or fall. If the winter is cold enough for the ground to freeze, start in the spring. The size of the bulblets will depend on the fertility of the soil and the weather conditions.

Here are ideas for several projects with garlic:

● Find a way to get the cloves to grow larger or faster, and you will have made both a scientific and economic contribution to the world.

● Some garlic varieties produce only five bulblets. Can you grow bulbs with more cloves or with bigger cloves using plant growth substances or other chemicals?

● Can you discover the ideal soil conditions for garlic?

● Are you able to produce a milder garlic? Or one that leaves no aftertaste?

● Can you invent a new spice based on garlic?

BE A CUTUP

The best way to grow most household plants is to start with cuttings. Cuttings have the same requirements as seeds: warmth, moisture, and a hospitable growth medium. Good media are plain sand or a mixture of sand, peat moss, and vermiculite or perlite. You are better off *not* using soil for cuttings unless you sterilize it first, since cuttings are not resistant to disease and garden soil contains bacteria and fungi. You can also root your cuttings in water. Although more cuttings are started in water than in any other rooting material, plant scientists and other experts don't use water because roots need air as well as water and roots grown in water tend to be more brittle.

How to Make a Cutting

Leaf cuttings are the easiest to make. Just remove a healthy leaf and its supporting *petiole* (the stalk that connects the leaf to the main stem). Insert the leaf cutting in the rooting medium inside of a small pot. Keep all leaf cuttings well watered. You can seal the cutting, pot and all, inside a plastic food bag. Then, one watering should be all that you need. Don't put the sealed cutting in a bright, sunny place or else the heat of the sun will cook it. If all goes well, in approximately three to four weeks roots will have formed (see Fig. 5.4).

Don't think you can only use leaf cuttings of house plants. You also can make stem cuttings (see Fig. 5.5) or root cuttings from woody plants or herbs (most common house plants are herbs), and from flowers and shrubs.

You may wonder how you can tell if the roots have formed. After all, you can't see inside the planting material. The easiest way is to *gently* tug the cutting. If the cutting easily slips out of the

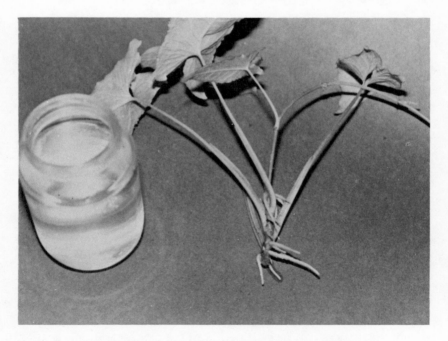

Figure 5.4. A variegated nephthytis cutting forming roots in water

Figure 5.5. Dieffenbachia. A stem cutting

medium, no roots have formed. Put the cutting back and try again the following week. If the cutting doesn't pull out after a gentle tug, let the roots develop undisturbed for another week. Then, transplant the cutting to its permanent home.

Root formation can be speeded up by using *auxins*, plant hormones that promote root growth. The auxin should be applied before insertion into the rooting material. Two commercial preparations that contain one or more auxins are Rootone and Transplantone. The concentration of plant hormones is very important. At high concentrations, for example, the chemical 2,4-D (2,4-dichlorophenoxyacetic acid) kills lawn weeds; at low concentrations, it speeds up weed growth. You can devise all sorts of projects and experiments using auxins.

Growth regulators work in strange ways when combined. Without auxins and cytokinins, for example, most cloning would be impossible. Auxins, cytokinins, and many other plant-growth-promoting substances can be obtained from the biological supply houses. Prices depend on quantity and concentration, and on availability.

Projects with Auxins

- At what concentration will 2,4-D fail to kill weeds?

- What is the highest concentration of 2,4-D that will stimulate plant growth? The lowest concentration?

- Compare the effects on plant growth of a natural auxin such as IAA (indole-3-acetic acid) and a synthetic hormone such as IBA (indole-3-butyric acid).

- What effect do auxins have on algae cells that live in water?

- What happens to cuttings if you add various auxins, natural and synthetic, in combination with a cytokinin (another type of plant growth substance) such as kinetin or our old friend gibberellic acid?

- Some known effects of auxins include causing leaves to fall and preventing potatoes from rotting. Can you discover any new effects that auxins have on plants, working alone or in combinations?

Table 5.1

PLANT REGULATORS

Regulator	Example	Effect
Auxin	IAA, 2, 4-D, IBA	Elongation of cells, high concentrations inhibit root growth; low concentrations promote root growth. Inhibits growth of lateral buds.
Gibberellin	Gibberellic acid	Very rapid stem growth.
Ethylene	Ethephon	Improves flavor and color of citrus fruits; ripening of fruits.
Cytokinin	Kinetin	Promotes rapid cell division.

GRAFTING

Mathematically, $1 + 1 = 2$, but plant scientists using a process called *grafting* can join the parts of two plants in such a way that the result is one new plant having the roots of one plant and the branches, leaves, and flowers of a different plant. The rooted part is called the *understock* (or *stock* for short); the other part is called the *scion*. The scion retains its own characteristics and gets only nourishment from the stock. For example, if the scion is a branch from a seedless orange tree and the stock is a lemon tree, seedless oranges will grow on the scion. The rest of the branches will produce lemons.

Grafting is usually done with woody plants, and the time to make your grafts is in spring, when the understocks are coming out of their "rest" or dormant stage. The scion should still be dormant too. Some plant scientists keep the scions dormant by refrigerating them until they are ready to be used. Citrus, pear, plum, peach, and apple trees are propagated by grafting. Other examples are the walnut and the mulberry tree.

In order for the graft to "take," the cambium (growth) layers of both must be in contact with each other. Another requirement is that the understock and the scion be closely related. If you like strawberries and bananas in your cereal, don't try to graft a strawberry stem onto a banana plant. It won't take! However, you *can* graft apples onto a pear tree or pears onto an apple tree. If you wish, lemons and limes can be grown on the same tree along with oranges or grapefruit. You also can do grafting with cacti (see Fig. 5.6).

How to Make a Graft

Refer to Figure 5.7.

1. To make the scion, use a sharp knife to remove a short piece of stem with one or two buds on it.

2. Cut into the understock so that the scion will "fit" and the cambium layers of the two plants touch.

3. Keep the cut surfaces from drying out and hold the scion in place using grafting tape and waterproof grafting wax.

Figure 5.6. A moon cactus grafted onto another cactus species

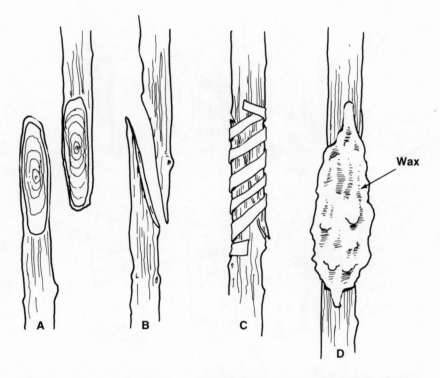

Figure 5.7. How to make a graft. (A) Make sloped cut through the bottom of the scion and through the top of the understock. (B) Cut the edges before joining them together. (C) Use grafting tape to hold the graft together. (D) Cover completed graft with wax.

How to Make a Bud Graft

Budding is a special type of grafting in which only a single bud is grafted onto the stock. This method is used for citrus fruits such as grapefruits, limes, and oranges and has proved successful with roses and avocados. Your chances of success are highest if the stock is young—that is, under two years of age—and if the bark of the scion is still dormant (see Fig. 5.8).

1. Use a sharp, thin knife or single-edged razor blade to remove the bud. Start about 1 cm below the bud and cut upward, through the bark. Include a thin sliver of wood behind the bark. End your cut approximately 1 cm above the bud. **(A–B, C)**

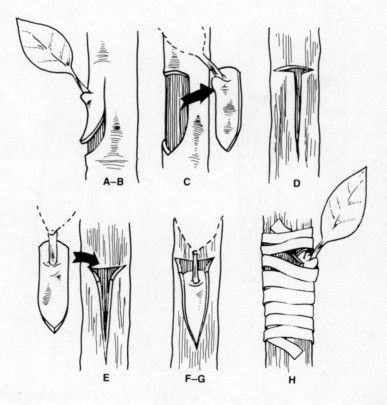

A–B   C   D

E   F–G   H

Figure 5.8. How to make a bud graft

2. Cut through the bark of the stock in the shape of a "T" where you want to insert the bud. Make the long cut 2 to 3 cm long and the cross cut 1 cm in length. **(D)**

3. Roll back the bark of the T-shaped cut. Make sure the cambium layer is exposed. The cambium will appear as a fine line between the bark and the wood as you look at the cut edge. **(E)**

4. Slip the bud piece under the flaps of the T cut, starting at the horizontal cut. **(F–G)**

5. If you did it correctly, the top of the bud piece will be below the cross cut of the T and the bud will stick out through the vertical cut. **(F–G)**

6. Tie the bud firmly in place using cut rubber bands as "string." Be sure the bud itself is exposed to the air. **(H)**

One reason for the high degree of success with this method is that the stock is only slightly injured. Again, care must be taken to keep the stock and bud from drying out. Buds are usually attached to the stock in midsummer, but growth won't start until the next spring.

Grafting: Odds and Ends

You may be wondering why plant breeders and researchers make grafts. Obviously this is a way to produce more branches that will yield seedless fruits. Also, some stocks cause the scions to be smaller than normal: Dwarf trees have become quite popular recently. Lastly, some grafts are more resistant to insects or soil disease as is the case with European grapes grafted onto American grape stocks.

I purposely have avoided going into detail concerning how to do whip, cleft, side, and other grafts. If you are interested in the details, consult some of the books in the Bibliography. One book you should read is *Plant Propagation* (Hartmann and Kester).

Household plants such as coleus and geraniums that have herbaceous (nonwoody) stems can also be grafted, but with much greater difficulty and less chance of success than woody plants such as fruit trees or grapevines. The reason is the herbaceous stems have less cambium per unit volume than woody stems. Also, these plants are never dormant and lose water faster and easier than trees. Herbaceous grafting can produce some unusual plants, for example, tomatoes growing from a potato plant. (See *Principles of Horticulture* listed in the Bibliography.)

DON'T LAUGH AT THE CLONES

Plant cell cloning has been going on for more than fifty years. Cloning or tissue culture consists of growing tiny pieces of plants, as small as the head of a pin, on a gelatinous mixture of nutrients. Adding plant hormones to the gelatin causes the small bits of tissue to grow into a lump of callus or scar tissue. This is similar to the scar tissue that seals a cut you may incur. Other added hormones change the scar or callus tissue from an undifferentiated (unspecialized) lump of tissue into a new plant with specialized tissues and organs, for example, conducting tissue and leaves.

Only since 1955 have scientists been able to stimulate undifferentiated cells to grow into whole plants. The secret is using a combination of cytokinin and auxin.

Advantages of Cloning

Cloning has revolutionized plant science in many ways. Cloned plants are usually free of disease. Thousands of rare, endangered plants can be grown quickly. Scientists can speedily evaluate plant chemicals and fertilizers. Valuable mutants can be discovered easily using cloned plants. Already, blight-resistant potatoes have been produced as well as disease-free strawberries and rice. Through the use of cloning, scientists hope to produce *super* vegetables and fruits.

Cloning is also important to the commercial florist. By growing plants in test tubes, nurserymen can increase their production by 500%. Instead of rooting cuttings in sand, they can now be cloned, a quicker and more successful process.

Other wonderful possibilities lie ahead for plant tissue culture. For example, hybrid plant cells that provide a balanced diet could be developed. At present, potatoes supply us mainly with starches; cloning could lead to potatoes that also give us our daily protein requirement. One approach to increasing the nutritional contents of plant cells is to produce hybrid plant cells by taking single cells of different plants (potato and tomato) and dissolving away their cell walls with enzymes. Then, get the "naked" cells to fuse into one cell that contains the inherited characteristics of both. At present, this type of research would be possible for you only in specialized plant research laboratories under the direction of a senior plant scientist.

To be successful, cloning must be carried out using sterile techniques in order to prevent bacterial infection in the early stages. In fact, for the beginning scientist the major problem is to maintain sterile conditions. Everything must be free of microorganisms, including plant parts, instruments, solutions, and glassware. You can maintain sterile conditions at home, but probably it is easier in a laboratory or in a school project room. Most school laboratories are equipped with sterilizers and autoclaves.

Basic Cloning Technique

It is wise to start off your work on plant tissue culture using a kit from a biological supply house. In addition to solutions and hardware, you will be provided with either carrot or lettuce seeds. After you acquire some experience this way, you can work on your own. Briefly, the procedure for cloning is as follows:

1. Bathe the seeds in a Clorox™ solution to kill any bacteria on their surface.

2. Rinse them several times in sterile water to get rid of the Clorox™.

3. Germinate them in sterile petri dishes on moistened sterile blotting paper. Both carrot and lettuce seeds will germinate quickly if the temperature is approximately 21° C.

4. When the seed germinates, remove a small piece of tissue using a sterile instrument, usually a scalpel. The small bit of tissue should be no larger than 0.5 cm. This is the *explant.*

5. Place the explant in another sterile petri dish, containing nutrients and an auxin, usually 2,4-D, which causes the callus to

form and enlarge. The formation of the callus takes from four to eight weeks at room temperature.

6. Remove under aseptic conditions small bits of the callus. Place them in test tubes containing a gelatin rich in nutrients and a mixture of auxin and cytokinin. The correct mixture will cause the mass of undifferentiated callus material to grow into the specialized cells of the roots, stems, and leaves of either carrot or lettuce plants. This may take another month or even six weeks.

Cloning Projects

Once you have mastered the basic techniques using the kit, you can experiment on your own. Here are a few project ideas.

• Can you speed up the process by formulating a better nutrient mix?

• Can you grow peony plants by this process?

• Which is better for growing Swedish ivy: cutting or cloning?

• Is cloning practical for trees and shrubs (privet, Norwegian spruce)?

• Can you devise a mix that produces more mutants than would be normally expected?

• What are the advantages of growing cloned cauliflower?

• Can squash or cucumber plants be cloned?

• Can you grow mutants that tolerate abnormally harsh environmental conditions such as sandy or salty soil? Such mutants would be commercially significant.

• Perhaps your mutants could utilize nitrogen from the air as well as from fertilizer or be suitable for nitrogen-fixing microorganisms to live in its roots.

6   LOOK, MA—
NO SOIL!

In every nation where the birth rate is larger than the death rate, the population of that country is increasing. Unfortunately, those countries that have the least ability to feed and care for their people have had the greatest population explosion. Although you and I might not suffer from hunger or malnutrition, this is not the case for many of the world's population. Can you think of nations in Africa and Asia where hunger is the rule rather than the exception? Worldwide, between 10 and 20 million starve to death each year and 2 billion are undernourished.

So far, scientists have been able to help farmers *produce* enough food for the steadily increasing population. This may sound strange in light of the last two sentences, but it is true! If we substituted the word *provide* for the word *produce,* then the sentence would have to be changed to read that we can't *provide* enough food for everyone.

Large quantities of food are lost due to spoilage during shipping and storage. Some countries, the United States, for example, have a surplus of food crops, while India and Cambodia may not produce enough for their citizens. Insects, rats, and plant diseases also cut down the amount of food available for humankind. One solution is to reduce our losses and distribute food differently. This sounds easy, but many political and economic problems are associated with this solution. We could also grow more food. However, most plant scientists believe we have reached the practical limit of growing crops on the land. It is too expensive or impractical to grow crops in deserts or on mountain tops.

Even tropical forest soils are not good for growing food crops. The main reason we cannot use more land areas for farming is topsoil lacking the nutrients plants require to grow and be healthy. Each year less and less farm land with adequate topsoil is available for supplying food to the people of the world. In the United States, shopping centers and homes are being built where potatoes and oranges once grew. Also in the United States, it is estimated that *3 billion tons* of topsoil are lost each year due to erosion.

One answer to the increasing population and the declining amount of farm land is to grow plants without soil, using dilute mineral solutions instead. This growing method is called *hydroponics.* Supporters of this method claim that with hydroponics, as opposed to ordinary soil agriculture, (1) plants receive just the right amount of water; (2) leaves can't be damaged by powdered fertilizers; (3) insect pests are eliminated; (4) weeds can't grow; (5) more plants can be grown closer together because the root systems are smaller; (6) growth is more rapid than in soil; (7) crop yields are larger; and (8) water and fertilizers are conserved because they can be reused.

Growing plants in nutrient water solutions does not guarantee success. Plants grown in hydroponic containers still need proper light, temperature, and humidity if they are to do well. Hydroponics has been used in research laboratories for over a hundred years, mainly in studies designed to investigate plant growth and development. Its recent commercial popularity began in 1936 when Professor William A. Gericke of the University of California held a press conference to show the tomato plants he had grown in hydroponic tanks. The plants were more than *7 meters (m)* tall, and he had to stand on a ladder to pick the tomatoes on top. Although hydroponics is still not widely used in the United States for commercial farming, Mexico, Israel, Japan, and Malaysia *do* have large-scale commercial hydroponic farming systems.

The first thing we will deal with is the nature of the nutrient solutions, since this is essential to hydroponics. Next, culture methods and special containers that you can buy or construct for growing your plants or seeds will be discussed. Then, you'll learn how to recognize and deal with deficiency conditions. And finally, you'll be presented with numerous ideas for experiments and projects.

NUTRIENT SOLUTIONS

Many formulas for nutrient solutions exist that can replace the minerals normally found in topsoil. The major-element formula given here works well in small quantities for many plants and supplies only the main elements plants require. You will need nitrogen, magnesium, potassium, calcium, phosphorus, and iron in the following form and quantities: 2.8 g sodium nitrate; 2.8 g superphosphate (16% P_2O_5); 1.4 g potassium nitrate; 1.4 g magnesium sulfate (Epsom salt); 0.4 g iron sulfate; 3.8 liters water. The only additional chemical you might need is dilute sulfuric acid in order to correct the pH of your nutrient solution.

Making Nutrient Solutions

Glass battery jars are best for making your nutrient solution. Dissolve all of the salts except for the iron sulfate in the 3.8 liters of water. The pH of the solution must be between 5.5 and 6 (see chapter 3). It is more likely that the pH of the solution will be high rather than low. If so, add a small amount of dilute sulfuric acid until the pH is within the recommended range. Now, add the iron sulfate and stir the mixture well. If you can, store your major-element solution in dark glass bottles or in clear bottles kept in the dark.

Although the trace elements of boron, manganese, copper, and zinc are often present as impurities in the chemical compounds used to make up our major-element solution, you want to play it safe by concocting and adding the following formula: 4 g manganese sulfate; 4 g boric acid crystals; 2 g zinc sulfate; 2 g copper sulfate.

Mix these salts thoroughly. Add 1 g of the dry mix and 20 ml of water to the major-element formula you previously made. You will now have 4 liters of solution rather than 3.8 liters. Store the dry, unused trace-element formula in a clean glass or plastic container that can be securely closed. Label the container clearly. You now have extra trace-element formula to add when you make up more nutrient solution. Stir the nutrient solution before adding it to plants growing in hydroponic containers. Change the nutrient solution once a month.

All of the chemical compounds in both formulas are readily available. They can be obtained from pharmacies, chemical supply companies, and scientific supply houses.

CULTURE MEDIUM

One of the best ways to anchor plants growing in hydroponic units is in sterile sand through which water and nutrients pass. This is called *sand culture*. In *aggregate culture*, vermiculite or gravel is used in place of sand. *Water culture* uses only a water solution of nutrients.

Water Culture

Water culture is the easiest method for you to use for growing *one plant* per container. However, water culture is impractical if you want to grow more than one plant per container. Philodendron and dieffenbachia do well without any special support system because their stems are quite stiff. Another plant that can easily be maintained hydroponically using the water culture method is the Chinese evergreen (*Aglaonema modestum*).

First, wash any glass container thoroughly with soap and hot water. Be sure to rinse well. Add a layer of washed gravel (available from pet stores) to the bottom of the container so that the roots can gain a foothold. Most plants need some sort of extra support, which water alone can't provide. Merely let the stem of the plants lean against the glass top of the container (see Fig. 6.1). Have only the roots and part of the stems under water, never the leaves! When handling the plants, take care not to injure the delicate roots.

Some important points to keep in mind when you are using the water culture method are:

1. Use tap water that has stood in an open jar for two days or more. This ensures that any chlorine gas dissolved in the water has escaped.

2. The pH of the water should be between 6 and 7. If you forgot what pH is, go back and read chapter 3.

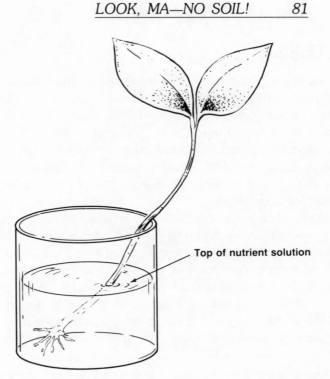

Top of nutrient solution

Figure 6.1. Water culture. The easiest method to use for one plant per container.

3. Change the nutrient solution (a water solution containing all the elements necessary for the growth and well-being of plants) every month. As the nutrient solution evaporates, *add only tap water* free of chlorine gas. This is because only the water evaporates; the nutrients remain in solution. Don't add any nutrient solution until it is time for the monthly change.

4. Algae growing on the plant roots or on the glass is not harmful; however, it may not look very nice.

5. Most plants require aeration of the roots once a day. Fish tank pumps make good aerating devices.

Other more complicated water culture methods can be found in most books on hydroponics (see Bibliography).

Sand Culture

Since water is not firm enough to support most plants, which lack fairly stiff stems, other materials are used for anchoring the roots in the nutrient solution container. Ferns are plants whose stems are too flexible to be supported by water alone, so silica sand (Fig. 6.2) is often used to support ferns and other such plants because it is cheap and easily obtained. Don't use sand composed of calcium salts, since this kind of sand will change the pH to 8, which is so high it will kill your plants.

Using sand or some other soilless medium presents the problem of getting the nutrient solution to the roots. Adding water from the top is the easiest but not the best way since you will have to add water quite frequently. One way to overcome this drawback is to use the "flowerpot-wick method." You will need a flowerpot, deep saucer or dish, sand, four cork or wooden blocks, and nylon or rayon wick 15 cm long (don't use cotton, wool, or linen).

Figure 6.3 shows you how to set up the sand-filled flowerpot.

Figure 6.2. Sand culture. Cuttings rooted in moist sand.

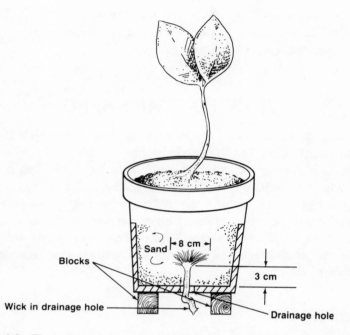

Figure 6.3. Flowerpot-wick method. Fill with sand or an aggregate mix.

Tease out the flowerpot end of the wick approximately 4 cm. Hold this end of the wick with one hand about 8 cm above the drainage hole. Add 3 to 5 cm of sand. Fan out the teased fibers. Add 3 more cm of sand. Now place the roots of the plant on top of sand. Fill the rest of the flowerpot with sand. The plant will be anchored and supported by the sand. Position the four blocks in the dish or saucer so that the flowerpot will be securely supported above the bottom of the dish. Pour the nutrient solution into the well. Remember to replace the nutrient solution only once a month. When the plant needs water at other times, add only tap water.

What to Grow in Sand Culture

Since your hydroponic setup is designed for a small-scale operation, stick to small plants you can grow from seeds or cuttings. Dwarf varieties will grow and flower nicely in your hydroponic setup. Vegetables and herbs that are easy to start from seed include cherry tomato, leaf lettuce, spinach, parsley, basil, and thyme. Recommended flower seeds to start with include zinnia, marigold, geranium, petunia, calceolaria, and pansy.

HOW TO START SEED

Start seeds in either a tray or individual containers, the only requirement being that the seed chamber is at least 8 cm deep. Fill the seed container with sand to a depth of 7 cm. Moisten the sand with water. Spread the seeds on top of the moist sand. If you are using a tray, leave room between seeds so that you can get the seedlings out for transplanting. Cover the seeds with no more than 1 cm of dry sand. Now, water *lightly*. Cover with plastic wrap until the seeds germinate. Place the seeds where the temperature is 18° C or higher.

Seedling Care

Growing plants without soil does not mean they can flower and thrive without *light, warmth, humidity,* and *clean air.* These requirements are for all plants and must be met, no matter where or how they are grown.

Once the seedlings have sprouted, remove the plastic wrap and move them into the light. Always keep the seeds moistened with the nutrient solution; make sure they do not dry out.

Moving to a Permanent Home

When the second pair of leaves appears, it is time to transplant the seedlings to their permanent home. Here is a safe procedure to follow:

1. Let the sand in the seed chamber dry out a little.

2. With one hand, place a spoon under the plant to be transplanted. At the same time, gently grasp a leaf firmly but carefully. This will help support the plant. Both hands are needed for this step.

3. Carefully remove the seedling, roots and all, by raising the spoon. A small ball of sand will come along with the roots. Leave the ball intact.

4. Put the seedling aside for a moment, resting in the spoon. Prepare a hole in the culture material large enough to receive the seedling.

5. Place the seedling in the hole. Remove the spoon.

6. Continue to hold the plant in place by a leaf. Using the other hand, cover the remaining space in the hole with the culture material.

CUTTINGS

The carnation is the ideal flower to grow hydroponically from a cutting. Here is how it is done:

1. Make the cuttings in November or December if you can, but, later in the winter is also okay.

2. Select healthy plants that are growing well and showing no sign of disease.

3. Use a sharp knife and make your cutting from the lower part of the flowering stem just below a node.

4. Make the cutting 8 to 12 cm long. Remove all the leaves from the base of the cutting. If you don't, the cutting will rot.

5. Root the cutting in moist sand in a tray or pot. Keep the sand moist with plain water while the roots are forming.

6. Once roots have formed, nourish the cutting with *half-strength* nutrient solution.

7. When the plants have five pairs of leaves, they can be nourished with a full-strength nutrient solution and moved to a permanent pot if they were started in a tray.

Roses, chrysanthemums, and hydrangeas are other flowers fre-quently grown without soil by means of cuttings. You might also try snake plants or coleus. For these, just follow the instructions given for the carnation.

AGGREGATE CULTURE

A mixture of gravel and perlite makes an excellent supporting material, even better than sand for plants growing in nutrient solutions. This mixture won't dry out too fast and provides good aeration. Do not use gravel alone, however, because it *does* dry out rapidly. Perlite is better than vermiculite because vermiculite often becomes waterlogged. A good mix is two parts perlite to one part gravel. You can also fill the container two-thirds full with perlite, then cover the perlite with enough gravel to fill the container.

Numerous other combinations of substances can be substituted for perlite and gravel. Here are a few possibilities:

5 parts gravel, 3 parts sand

2 parts vermiculite, 1 part sand

Styrofoam and crushed brick (no ratios are given because the size of both can vary greatly; a good project would be to find the ideal size and ratio of each)

You can fill a flowerpot (such as the one illustrated in Figure 6.2) with an aggregate mix instead of sand and your hydroponic setup will function just as well, possibly better! Why not investigate which supporting material is the best. Use a water culture unit as your control.

TIMED HYDROPONIC UNIT

You can buy hydroponic kits from biological supply houses, gardening establishments, greenhouse equipment companies, and other sources. The kits contain growing chambers, pumps, plants, and nutrient solutions.

If you are mechanically talented, you might want to make a hydroponic unit that feeds your plants at timed intervals. A pump forces air into the nutrient chamber. The air pressure then forces the nutrient solution into the upper container, which holds the soil-less medium. After a set number of hours, the timer shuts off the pump and the nutrient solution drains back into the lower container by gravity. The basic unit is made with two styrofoam coolers, one that can sit inside of the other, with an air space between. Other necessary supplies are plastic tubing, an electric timer, and an aquarium air pump, a small piece of fiberglass screening (5 by 5 cm), and silicone sealant. Figure 6.4 shows you the basic plan. Measurements will vary with the size of the styrofoam containers used. Here's how to do it.

1. Cut the hole in the center of the upper plastic chamber.

2. Set the upper chamber inside of the lower unit. Measure the height of the space available for the nutrient solution and air.

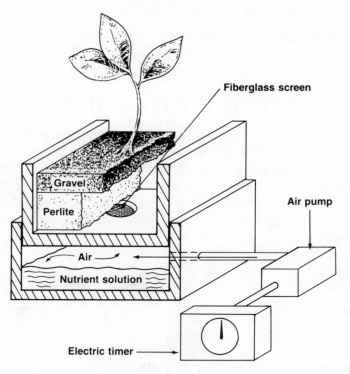

Figure 6.4. Homemade hydroponic unit is best if you are growing many plants hydroponically.

3. Cut the hole for the tubing to enter the lower chamber. Be sure the tubing fits snugly.

4. Seat the upper unit into the lower, cementing and sealing it in place with silicone sealant.

5. Insert the tubing into the hole you made in the lower unit. Seal the tubing in place with silicone. The fitting must be air- and watertight.

6. Cement the fiberglass screen over the opening in the base of the upper unit.

7. Fill the reservoir with the nutrient solution.

8. Add perlite, the plants, and gravel to the upper unit as shown in Figure 6.4.

9. Connect the pump and timer as shown.

10. Set the timer so the pump works two hours a day, preferably in the morning.

DEFICIENCY CONDITIONS

Your plants will react in a characteristic manner if they get insufficient essential elements. *Malnutrition* stems from many causes, not necessarily a faulty nutrient mix. The culture material may be the problem. For example, you may have used vermiculite instead of perlite, and the vermiculite became waterlogged. The chart that follows lists the essential elements and the corresponding deficiency symptoms.

Element	*Deficiency Condition*
Nitrogen	Large root system; tiny stems and leaves
Potassium	Yellowing of leaf borders
Phosphorus	Plant fails to grow to normal size
Magnesium	Flowering is delayed
Calcium	Very small roots
Iron	Leaf tips at top of plant lose green color

Experiments and Projects in Hydroponics

● What is the effect on plant growth of suspending a piece of mesh such as cheesecloth on top of the nutrient material?

● Does tobacco smoke carry viruses that attack plants growing in small hydroponic units?

● Can dwarf trees be grown hydroponically? Can you devise a special support system for them?

● Are the claims made for hydroponics true: Are there fewer insect pests, more plants grown in less space, fewer diseased plants? Is less fertilizer used and the yield really higher?

● Can plants be grown from bulbs and corms in hydroponic units?

● Can you devise a special nutrient formula for a particular plant?

● Can you grow corn from seed hydroponically?

● Can cacti be grown by using the water culture technique?

● How does nutrient solution affect the growth of such vegetables as the sweet potato, carrot, or rutabaga? Since these vegetables will grow in plain water, you have a means for comparison and control.

● Repeat the above experiment using avocado seeds or pineapple tops.

● Can hydroponics speed up growth so that semitropical plants can be grown out-of-doors in a temperate climate zone? Can you grow bananas, mangos, or kiwis in cold climate locations?

● How much water will a tomato plant (or a plant of your choice) absorb in 24 hours?

● What is the effect of lowering/raising the pH of the nutrient solution on the growth of carnations/cucumbers?

● Can treated sewer water be used to make up hydroponic solutions?

• What concentration of salt is lethal (deadly) to hydroponic tomato plants?

• What concentration of iron or other trace mineral is toxic (poisonous) to celery plants growing in hydroponic tanks?

• One of the reasons why hydroponics is not widespread in the United States is because it is expensive. A nice project would be to work out a cost analysis and determine at what cost it is cheaper to grow tomatoes by using hydroponics rather than on 2 sq m of soil. Be sure to figure the cost of adding compost, manure, and fertilizer to that small plot of poor soil. Local colleges that have agriculture courses or the U.S. Department of Agriculture can provide you with help in determining costs and setting up this project.

7   GENETICS: FROM YESTERDAY TO TOMORROW

YESTERDAY'S GENETICS

People have been improving plants for thousands of years by crossbreeding and selection. The aim is to grow more and better food for the world's population. Why some attempts were successful and others not was not understood until Gregor Mendel demonstrated around 1860 that structures in the sperm and egg cells are responsible for inherited traits. Later on, these structures were given the name *genes*. In each generation these structures are combined in a different way. Mendel was successful where others failed because he incorporated mathematics into his carefully planned crossbreeding experiments.

Mendel was always concerned with improving crops. At an early age, he developed new varieties of fruits and vegetables. However, he is best known for his genetic experiments with garden pea plants. His decision to use garden peas was not an accident: Peas were available, were easy to grow, and a rapid producer of flowers. More important, peas have traits that are easily seen, such as seed color. Finally, the flowers normally self-pollinate themselves. Soon after Mendel graduated from the University of Vienna, he raised over thirty "pure" strains of garden peas in the Austrian monastery where he lived and worked as a monk. These were used in his genetic experiments.

Some scientists who have repeated Mendel's experiments report that their results do not agree as closely with Mendel's mathematically expected frequencies. This might be a starting place for your genetic experiments, namely, to repeat some of Gregor Mendel's classic experiments and see how *your* results compare with his.

Mendel's Experiments

You can start with dwarf and normal-size pea seeds, which can be bought in seed stores, in gardening shops, and through biological supply houses. If you prefer to experiment with other traits such as wrinkled and smooth seeds, or green and yellow seeds, then your best sources of seeds are the biological supply companies.

Using seeds of contrasting traits (dwarf and normal size), you can experimentally check the accuracy of two of Mendel's three laws.

1. *Law of Dominance.* When two pure pea plants with contrasting traits (tall/dwarf) are crossed and meet in an offspring, one trait will mask the other. The trait that appears is the *dominant* trait and the hidden trait is termed *recessive*. The offspring in which they meet is a *hybrid*. One hundred percent of the hybrids will show the dominant trait.

2. *Law of Segregation.* When hybrids are crossed, there is a sorting out of characteristics: 75% of the offspring of the two hybrids will show the dominant trait, and 25% will show the recessive trait. However, of those that show the dominant trait, 25% of the total offspring will be pure for that trait and 50% will be hybrid.

3. *Law of Independent Assortment.* You may decide to check the results of Mendel's third law, which deals with dihybrids (hybrid for two traits). This experiment is a bit more complex since you have to work with two pairs of traits at once. This demands accurate observations and record keeping. Instead of the offspring having one of three possible traits to look for when hybrids are crossed, there are *sixteen* possibilities in the case of a dihybrid cross. An offspring could be pure dominant for both traits, hybrid for one trait

and recessive for the other, and so on. You must start with seeds that are pure for the contrasting traits. You could use seeds that produce dwarf pea plants with green seeds and normal-size pea plants bearing yellow seeds. Your only reliable source for these seeds are the biological supply companies.

If dihybrid pea plants are crossed, each trait will be inherited independent of the other. For example, the trait for seed color will not be affected by the manner in which the trait for the size of the plant is inherited. In the offspring, the traits will appear in the ratio of 9:3:3:1. This means ideally that if there are 16 offspring, 9 will show both dominant traits (normal size and yellow seeds) and 3 will show one dominant trait and one recessive trait (normal size and green seeds or dwarf plant with yellow seeds). The second "3" will show the other dominant and recessive traits, while only 1 will show both recessive traits (dwarf plant with green seeds).

Growing Peas

It is best to start your garden pea seeds outside in the cool weather. Peas don't grow very well in the hot summer months and they die in the winter. However, they *can* be grown in the house (especially the dwarf varieties), provided they are given enough soil and light (16 hours). Make sure the soil has lots of humus and that its pH is between 5.5 and 7.

Pea seeds don't transplant well, so start them where you want them to grow. Plant them about 2 cm below the surface of the soil and keep the soil well watered. The seeds will sprout in approximately two weeks in the garden and sooner indoors. As they get taller, you will need to support them in some way; wooden stakes are easy to find and use. Keep in mind that dwarf peas will flower about two weeks earlier than the normal tall peas. If you plan on crossing these peas, start the tall ones two weeks before you plant the dwarf pea seeds.

The pea flower, as you can see in Figures 7.1, 7.2, and 7.3, is designed for self-pollination. The male (anther) and female (pistil) reproductive parts are entirely covered by the large petals. This flower is unusual because it does not open until after fertilization has taken place. What Mendel did and what you must do is to artificially cross-pollinate the flowers. If you wanted to cross a dwarf with a normal-size plant, use the following procedure:

1. Carefully open an immature flower bud on the tall plant by prying open or cutting the petals.
2. Remove the anthers with forceps or tweezers (consult Fig. 7.2).
3. Cover the flower with a paper bag tied in place to insure that no stray pollen lands on the stigma.
4. Wait until the flower matures before attempting to cross-pollinate it.
5. The flowers of a dwarf plant should mature at the same time if you planted on schedule. Open a dwarf flower as you did the tall flower.
6. Using a fine brush, brush the anthers in order to collect the pollen from the dwarf (consult Fig. 7.3).
7. Brush this pollen onto the stigma of the tall plant's flower.
8. Re-cover the cross-pollinated flower with the paper bag.
9. Collect the seeds when the pods ripen. These are your hybrids. Ripe pods will be twisted and dry. Sometimes, the ripened pods may break open, exposing the seeds. Don't pick the peas too early: you are going to plant them, not eat them!

Figure 7.1. Model of pea flower showing the five petals (keel, equals two petals fused together, two wings, and one large banner).

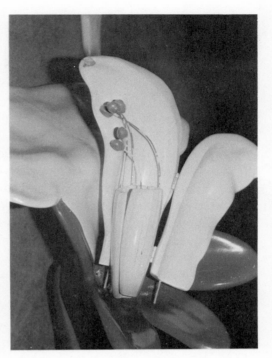

Figure 7.2. One wing removed from the pea flower model revealing stamens that have globular anthers on top of supporting filaments.

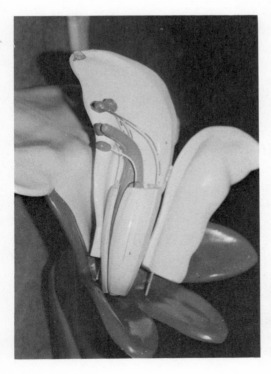

Figure 7.3. Pistil in place on pea flower model. The pistil is between the stamens; the stigma and style are clearly visible.

If asked "Which would most likely be the better, healthier plant: a pure breed or a hybrid?" most people unfamiliar with the principles of genetics would select the pure breed. Geneticists would pick the hybrid since most hybrids are stronger and healthier. Experiments and experience have shown that, usually, hybrids (1) are better able to resist changes in the environment than pure varieties; (2) resist infections and pests better than pure lines; and (3) combine the best traits of both parents. In 1900, several varieties of red winter wheat from Russia were introduced into the United States. Much of the wheat varieties grown in the United States today were developed from the hybrid offspring of the Russian and U.S. wheat developed over 80 years ago.

When you are trying to produce hybrids keep the following points in mind:

- The two species must be closely related.

- Use paper bags to prevent stray pollen from getting onto the stigmas.

- Some offspring produced by cross-pollination are sterile.

Now that you know the technique (outlined above for crossing pea plants), you can try to hybridize any number of plants. Morning-glory is a good species to cross-pollinate. Sorghum is another. They can be forced to flower in about 2 months provided they are exposed to a "short day," that is, 16 hours of darkness.

Projects with Hybrids

- Can you produce a better morning-glory hybrid?

- A triple cross between a tangerine, an orange, and a grapefruit has been successful. Can you cross two of these citrus fruits with success?

- Try to produce a rose bush that has silver-colored roses by cross-pollinating several strains of roses.

GENETIC ERRORS

Everybody makes errors, some serious, others not. If a printer accidentally changed a sentence in a book, you might be unable to understand what that sentence meant. This could simply be annoying or it could be serious: What if instructions for building a house were garbled?

Errors also occur in genes, but most of the time the result is a trait that is bad for the plant, animal, or person which it occurs in. Furthermore, that gene error will be passed on to the offspring and future generations. A genetic change is called a *mutation*. One example of a mutation is seedless grapes. Lacking seeds may be good for the people who hate grapes with seeds, but a grape plant lacking seeds cannot reproduce sexually.

Plant researchers are always on the lookout for a useful mutation, a disease-resistant plant. Plant scientists also experiment with *mutagens* (substances that cause mutations) in the hopes of producing a useful plant mutation. You can do the same, with some limitations, because many mutagens have harmful effects on people as well as on plants.

Mutation Experiments

X-rays were the first mutagen to be discovered. In 1927 at Columbia University, Dr. H. J. Muller exposed fruit flies to strong doses of X-rays. The radiated flies were mated with flies that were not X-rayed. The offspring had many more mutations than expected—150% more! All kinds of mutations were seen. Some mutants had small wings or no wings, others had poorly formed eyes, and some had no eyes, and so on. There was no order to the mutations. Muller could not produce a mutation "to order," something scientists still cannot do with radiation. X-ray mutations occur randomly.

Since the only places that have X-ray machines are research laboratories, hospitals, and doctors' and dentists' offices, you will have to ask someone who works in one of these places, and who is authorized to use the equipment, to expose your plants or seeds to X-rays for you. Only licensed technicians or qualified physicians may operate X-ray machines. To make your experimental data more meaningful, ask how much radiation your experimental flowers or seeds received. X-rays from medical or dental X-ray units can be varied in three ways. They are (1) *distance,* between the X-ray tube and the plants; (2) *milliampere seconds,* a measure of the strength of the X-rays; and (3) *kilovolts,* a measure of the penetrating power of the X-rays.

Divide your seeds into five groups. Ask that all the groups be exposed to X-rays for 50 milliampere seconds at 100 kilovolts. Place one group of seeds 25 cm away from the X-ray source, the next group 50 cm away, the third group 75 cm away, and the fourth group 1 m away. The fifth group will be the control group and will receive no radiation at all. Another possibility is to vary the voltage and keep the distance and milliamperes constant. X-rays of 25, 50, 75, and 100 kilovolts for 10 milliampere seconds and 1 m distance could also be used for the four experimental groups.

Good plants to X-ray are those that can be grown indoors or outside with relative ease. Examples include cosmos, marigold, dwarf tomatoes, and leaf lettuce. Compare the plants that grow from irradiated seeds with plants grown from seeds not exposed to X-rays. Be sure to carry on this experiment for *two* generations.

Cross-pollinate two flowers exposed to X-rays, and do the same with your control plants. Compare your experimental and control plants. Look for mutants with different shaped or colored leaves. Dwarf plants, tall plants, or plants that lack flowers are possible mutations that you might discover.

Radiation from radioactive elements produces similar effects to those of X-rays. Again, it may be possible to have your science teacher or a scientist at a university laboratory help you. Because radiation is so dangerous you won't be able to obtain your own sources. You will have to obtain irradiated seeds from a scientific supply company if you don't have access to a laboratory researcher. Handling irradiated seeds is *not* dangerous, however. The experiments you can do with irradiated seeds are similar to those outlined for X-rays. Here are a few possibilities:

• Compare the germination rate (percentage of seeds that germinate) of irradiated seeds with a control group.

• Expose seeds of different species to radioactivity. Which seed's germination rate is affected the most? Which is affected the least?

• How many seeds grow into abnormal plants that are stunted or fail to produce flowers?

Ultraviolet light causes mutations in bacteria. Try exposing flowers and seeds to varying amounts of ultraviolet light. Then, observe them and their offspring for mutations. *Do this with the assistance, and under the supervision, of an experienced adult. Care must be taken that no one gets exposed to this harmful radiation, since ultraviolet light is known to cause skin cancer.*

Certain chemicals are mutagens. Experiment with the seeds described earlier in the section on X-rays. Try varying concentrations of nitrous acid, organophosphates, and colchicine.

Nitrous acid is used to make sodium nitrite, a preservative for hot dogs, bacon, and other preserved meats that causes cancer in animals and may be hazardous to humans as well. Use caution and work under supervision when handling this chemical.

Organophosphates are chemical compounds that are the active ingredients in a number of insecticides. Check the labels of insecticide cans to discover if this ingredient is present. ("Phosphate" will be part of the chemical name.) If so, you could do two experiments at the same time: (1) Determine its effectiveness as an insecticide. (2) Determine its mutagenic effect.

Colchicine is a compound that doubles the chromosome number of plant cells and prevents cell division from occurring. Chromosomes are nuclear structures composed of genes. One result of having an extra set of chromosomes is larger flowers and fruits. Another result is usually sterile flowers become fertile (able to reproduce).

Colchicine has been used in very dilute amounts to treat people suffering from gout. In larger amounts, it has many harmful effects on humans, attacking the nervous system as well as other body parts. Death can result from swallowing as little as 0.02 g. *Colchicine is absorbed easily through the skin and never should be handled unless gloves are worn.*

Plant scientists have used colchicine to double the size of a species of grape that is tasty and grows well in the southern United States but cannot be marketed because of its small size. A dash of colchicine on some of the grape seeds made them grow into grapes large enough to be marketed. The grapes became another source of income for the farmers and another food for all of us. Researchers also have used colchicine to change a hybrid wheat plant with many fine qualities from a sterile hybrid to a fertile one. The offspring were reproduced in quantity and now are being grown in the United States and other countries.

Weak solutions of this chemical, no stronger than 1%, are applied to flower buds and seeds for experimental purposes. Because of the danger, only qualified persons should handle colchicine. *You should not!*

Safer substitutes for colchicine are chloral hydrate and sodium deoxycholate. But don't forget: *Handle all chemicals with care and caution.* Practically all scientific supply companies stock and sell these chemicals.

Cosmos and zinnias are excellent flowers and seeds to treat with colchicine and its substitutes.

PLANT GENETICS—TODAY AND TOMORROW

The pea plant was a favorite of scientists like Mendel who worked in the nineteenth century, but by the 1930s the fruit fly had become more popular as an experimental organism. Today, scientists prefer working with bacteria. Plant geneticists are concentrating upon the *Agrobacterium tumefaciens,* the soil bacterium that causes *crown gall* disease in many plants, including grapes, tobacco, and peaches.

Galls are small, brown *tumors* (a tumor is a mass of cells that grows without limits and performs no useful function). The crown is the region of the plant where the stem and the root meet.

Geneticists are interested in this bacterium because it inserts foreign genes (its own) into green plants. Infected plant cells follow the instructions of the foreign, injected genes and form a gall. *Agrobacterium* reproduce in great numbers inside of the gall.

Geneticists want to insert foreign genes into plants and have the plants' cells follow those genes' instructions. This process is called *genetic engineering.* Geneticists don't want the plants to form galls, but they do want plants to produce substances that make the plants immune to certain diseases or enable the plants to use fertilizer more efficiently.

Genetic engineering may seem to you to accomplish exactly what now is being achieved by cross-pollination. You are correct! Unfortunately, cross-pollination takes place only between the same or similar plant species. Producing a hybrid possessing just one new useful trait may take years and involve many crosses before success is achieved. Genetic engineering is a relatively quick process. New genes can be introduced in a few weeks or less. Most important, the new genes need not even come from a plant! All sorts of strange genes can be introduced. One hope is to incorporate the genes of nitrogen-fixing bacteria into corn plants. Then these corn plants will be able to use the nitrogen in the air to make proteins, instead of requiring the nitrogen that is the main ingredient in fertilizer. When this is finally done, farmers will not have to add such expensive fertilizer to their cornfields.

Genetic engineering combined with cloning (see chapter 5) can lead to the quick production of new and useful hybrids in quantity. We can clone some plants with ease, but this is not the case with genetic engineering.

The Genetic Engineering Process

A method of insertion is required in order for scientists to insert new genes in order to produce new varieties of plants. So far, the only *efficient* way scientists have found is to use *Agrobacterium tumefaciens* as the inserting agent. To understand this procedure, some knowledge of genetic chemistry is necessary.

All living things, except for viruses, contain one or more chromosomes composed of DNA, the chemical that makes up genes. Bacteria also contain a *plasmid* (a ring made of DNA near the chromosome). The DNA of the chromosome contains the information for the bacteria's life activities—reproduction, excretion, etc. The plasmid's DNA contains information for special traits, including the ability to infect plants and resistance to antibiotics.

To engineer a trait into a plant using *A. tumefaciens* requires the following steps, which are also illustrated in Figure 7.4a–h:

1. The plasmid is removed from the *Agrobacterium*. It is opened and a piece of "passenger DNA" inserted. The passenger DNA is the gene you want the plant to receive. Now, the ring is closed. The plasmid has been *engineered* (Fig. 7.4a).

2. The engineered plasmid containing the desirable passenger DNA (gene) is put into normal *A. tumefaciens*. The *Agrobacterium* now has *two* plasmids (Fig. 7.4b).

3. The bacteria are reproduced in great numbers. There is an exchange of DNA between the inserted plasmid and the natural one. This happens because of the process called "recombination." The passenger DNA fragment combines with the normal plasmid. Now, the gene on the normal plasmid that causes crown gall disease recombines with the inserted plasmid. For both of these events to take place is rare, but it does happen! This rare but essential event is called *double crossover recombination* (Figs. 7.4c, 7.4d).

4. Scientists search for bacteria whose normal plasmid now contains the passenger DNA but not the gene for crown gall disease. The plasmid that now contains the DNA for crown gall disease is ejected from the bacterium (Fig. 7.4d).

5. A plant cell is infected with the bacterium containing the plasmid with the passenger DNA in it (Fig. 7.4e).

6. The engineered plasmid becomes attached to the DNA of a chromosome in the plant cell nucleus (Fig. 7.4f).

7. The plant cell is cloned by tissue culture techniques (Fig. 7.4g).

8. The clones grow into healthy plants. All the cloned plants possess the passenger DNA and show that trait (Fig. 7.4h).

What you have just read is a simplified account of the basics of genetic engineering. Unfortunately, genetic engineering cannot be done at home or anywhere outside of the most up-to-date research laboratory. Furthermore, many scientists, each an expert in one phase of this elaborate process, must cooperate in order that a genetic engineering project be successful. You might be able to become involved in genetic engineering experiments as part of a summer institute for high-school students given at a college or university.

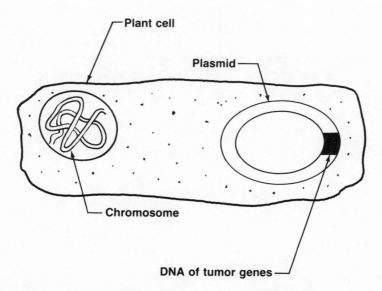

Plant cell

Plasmid

Chromosome

DNA of tumor genes

Figure 7.4.a. How to engineer a new gene into plants. Normal agrobacterium tumefaciens (before engineering)

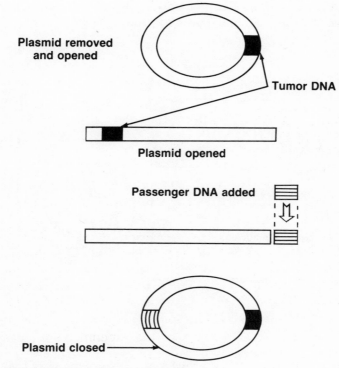

Plasmid removed and opened

Tumor DNA

Plasmid opened

Passenger DNA added

Plasmid closed

Figure 7.4.b. Engineering the plasmid

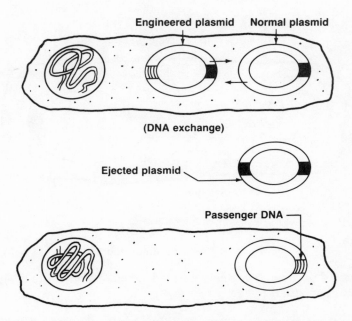

Figure 7.4.c. and d. (top) Recombination. *(bottom)* The plasmid contains the passenger DNA but not the gene for crown gall disease.

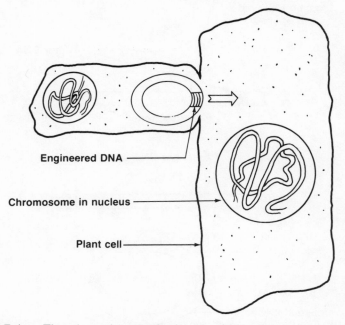

Figure 7.4.e. The plasmid is attached to the DNA of the plant cell.

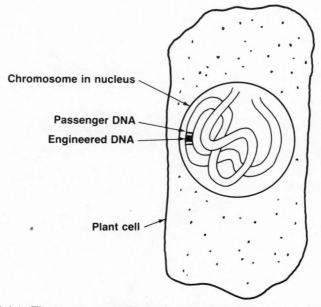

Chromosome in nucleus

Passenger DNA

Engineered DNA

Plant cell

Figure 7.4.f. The engineered DNA incorporated into the plant chromosome

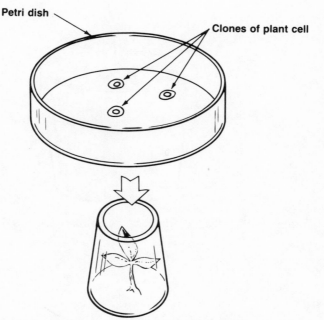

Petri dish

Clones of plant cell

Figure 7.4.g. Tissue culture techniques are used to clone the plant cell.

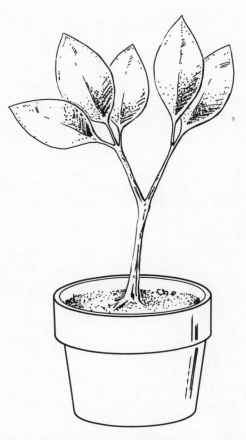

Figure 7.4.h. Healthy adult plant with the engineered trait

PROBLEMS TO BE SOLVED TOMORROW

Extend the range of applications in genetic engineering. Genetic engineering is limited to those plants that A. *tumefaciens* infects (petunia, tomato, and tobacco). This bacterium will not infect grasses, cereals, corn, or most of the plants that we grow for food.

• Increase the rate of success for inserting engineered plasmids, currently very low.

• Discover the genes for resistance to pesticides and weed killers: Only genes whose traits are known can be successfully engineered.

• Learn much more about plant genetics, a very complex subject because some plant cells contain eight times the amount of DNA found in a cell in your body!

• Successfully clone peanuts, soybeans, wheat, and other cereals. This is a possible experiment for you to design and carry out (see chapter 5).

8   PLANT PEST EXPERIMENTS

We have a lot of nonhuman competition for the food we grow. Our main competitors are not birds and mammals but the many insect pests that feed on plants. Just to give you some idea of what insects can do, consider this: Gypsy moths can feed on 500 kinds of trees and vines, including apple and cherry trees. Many of the plants the gypsy moth won't eat, the Japanese beetle will. This beetle defoliates (removes the leaves) on more than 270 kinds of flowering plants and trees. Both of these insect pests can be found in great numbers in the northeastern United States and are spreading to the southern and western states.

In this chapter we are going to experiment with and explore ways to prevent insects from eating our crops before we do. In some cases the experiments are designed to kill these pests. Other experiments will explore methods of deterring the insects. Both kinds of experiments are significant and important because of the need to feed the ever-increasing world population we spoke about in an earlier chapter. A second concern is the fact that the amount of land suitable for agriculture cannot be increased and is actually decreasing. We can't waste our food feeding insects. The challenge is significant because in the process of wiping out or deterring insect pests, we don't want to harm the environment for fish, birds, or mammals, including ourselves.

Billions of dollars' worth of food crops are lost each year to insect pests. These losses are not new. References to locusts devouring crops can be found in the Bible. Our most effective weapon in the war against harmful insects is the use of chemical poisons (insecticides). Unfortunately, in recent years we have had a number of serious problems with chemical insecticides, including the deaths of people and birds. This is why scientists are searching for new ways to repel and kill the harmful bugs.

The difficulty with insect poisons is that they have no way of knowing if what they are killing is helpful or harmful. We learned about the undesirable effects of insecticides the hard way. Pelican, eagle, falcon, and other bird populations were lowering to the point where extinction was almost certain. Scientists discovered that the shells of these birds were so thin and fragile that they broke before the chicks could hatch. DDT, an insecticide that was widely used against many insect pests, was the cause. Birds that ate insects poisoned with DDT laid eggs with extremely thin shells. DDT, in small doses, is harmless to animals except for insects. Unfortunately, DDT does not break down in animal or bird tissues, but remains unchanged. Small amounts taken over a long time accumulate, and numerous fish, birds, and mammals have died from DDT poisoning. The result is a total ban on DDT in the United States.

COMMERCIAL BUG KILLERS

I am *against* experimenting with chemical insecticides (see Fig. 8.1). All insect poisons are tested by the company before they are released for sale, so there is little point in experimenting to see if they kill bugs. If you *do* decide to work with a commercial insecticide, *read the label carefully!* Most labels caution against spraying on people or animals; many on foods or dishes. Still others warn against eating fruits or vegetables for a week or two after spraying. An effective insecticide will remain on a plant in active form for a week or longer. In this way, any insect that comes for a meal several days after you spray will be killed. You could test two or three insecticides to see which is best for killing aphids or tomato cutworms, but you will learn more by trying one of the alternative experiments or projects described on the pages that follow.

INSECT PREDATORS

Not all insects are pests; some insects feed on destructive insects without disturbing plants. When properly used, they are effective and safe in controlling insect pests. Furthermore, their prey don't

Figure 8.1. Malathion® is effective against aphids, spider mites, and mealy bugs.

become immune to their insect predators. The same can't be said for insecticides. Despite intensive spraying with chemicals, more than 400 insect pest species have become resistant to one or more chemical insecticide.

The praying mantis is one such useful insect that you probably have seen. When young, this predator can eat many times its own weight in harmful aphids daily. Adults feed on large numbers of caterpillars, grasshoppers, and other harmful insects each day.

Projects with Predators

Here are some ideas for projects with insectivorous (insect eating) insects and spiders. (Spiders are not insects, but are arachnids.)

1. Raise spiders near your plants. Spiders of all sorts will be found in your garden provided there are sufficient insects of all sorts

for them to eat. If you have sprayed insecticides in your garden plot, don't search for spiders. The chemicals will have killed them and their food supply, including helpful insects that pollinate the flowers.

● What insects do they eliminate?

● How effective are their webs as insect traps? Which insect species become trapped in the webs? How many insects are caught in a day? In a week?

● How do web shape, design (different species of spiders spin different shaped webs), and size contribute to its effectiveness?

● If you spray a spider's web with a floral scent (air freshener, perfume, etc.), will the web trap more insects? Which scent works best?

● What proportion of helpful insects are trapped in spider's webs?

2. Ladybugs, praying mantises, and some kinds of wasps prey upon aphids, Japanese beetles, and other insect pests. You can purchase them from natural pest control companies or some of the large gardening firms. Two sources are Natural Pest Controls, 9397 Premier Way, Sacramento, CA 95826, and W. A. Burpee Co., 300 Park Avenue, Warminister, PA 18974. Adult ladybugs, praying mantis egg cases (each containing 100 to 200 eggs), and the eggs of parasitic Trichogramma wasps are sent.

Any of these three predatory insects are excellent choices for controlling insect pests in your garden. However, you must be sure there is ample food for the helpful insects or they will move away in search of a good meal! These predators like other foods, including pollen and nectar from flowers. Planting many different vegetables and flowers will encourage your purchased predatory insects to stay in your yard or garden because the chance of planting a food source they like will be improved.

● What flowers do these helpful insects prefer as another food source?

● Do herbs such as chives and mint repel ladybugs and other desirable insects?

• Which insect pests do mantises prefer? Which insect pests do they ignore?

• How many predators (ladybugs) are the ideal number for a 10 sq m garden plot?

ORGANIC INSECTICIDES

Some insecticides made from plants that are harmless to humans, other mammals, and birds are termed *organic*. Chemical insecticides are *synthetic* since they are made in chemical laboratories. Pyrethrum and rotenone are two common organic insecticides recommended for use against Japanese beetles and aphids. Another, ryania, is poisonous to corn borers. Rotenone and ryania, which are derived from South American plants, can be purchased in seed and garden shops.

Here are a few questions to get you started with projects with organic pesticides:

• How do organic insecticides compare in effectiveness with synthetic chemical insecticides?

• How long do organic insecticides remain effective?

• What effect do these substances have on earthworms? On helpful insects (ladybugs and praying mantises) and spiders?

HOME-BREWED INSECTICIDES

In our search for insecticides that will adversely affect only the insect pests, we should start at home. In the same way that some of you don't like lemons or a lot of pepper, pests in the insect world hate the odor or taste of some common nonpoisonous substances. Our task is to find these substances and learn how to apply them and in what quantities.

1. Stale beer in a small dish lures snails and slugs. When they crawl into the dish, they drown in the beer!

• What is the attractant in stale beer: the odor, the taste, or the color? What is the best size container to use? Which brand of beer is most effective? Will fresh beer straight from the can work as well as beer that has been exposed to the air for four hours or more?

• Will other alcoholic beverages work as well or better? For a start, experiment with wine of known alcoholic content.

2. Many house plants are attacked by spider mites, which feed on their juices. Their leaves lose their green color and look as if they are made of bronze. Spider mites are quite small, and you must look hard to find them. A spray made from 53 g of wheat flour stirred with 2 ml of buttermilk and 98 ml of water is said to be effective against these bugs. An empty spray bottle that once contained window cleaner will make an ideal sprayer if rinsed a few times with hot water. Design experiments to answer one or more of these questions:

• Does this spray kill the mites, their eggs, or both?

• How long is the spray effective?

• Are house plants harmed in any way by the spray?

• Do people find the spray objectionable?

• Are houseflies attracted to sprayed plants?

• Can other insect pests be killed with this spray?

3. What follows is a "recipe" for a homemade insecticide. A recipe is less precise than a formula.

Garlic juice from 1 clove of garlic.

Onion juice from 1 medium onion cut into fine pieces, then crushed.

Tobacco juice made from allowing one cigarette to remain overnight in a glass of water. Discard cigarette.

1 teaspoon hot pepper powder

1 teaspoon oregano

1 teaspoon liquid soap (don't use detergent—it's not organic!)

Mix ingredients in two liters of water. Stir well. Pour mixture into a jar. Cap tightly. Allow the solution to stand three days before using.

Try to answer the following using an experiment of your own design:

• What house plant insects (spider mites, white flies, mealybugs) will the mixture kill?

• Does it kill or repel tomato hornworms? Squash borers? Aphids?

• If the recipe shows promise, can you devise a formula rather than an inaccurate recipe?

• How long does the insecticide spray remain effective?

• If used on garden plants, must they be resprayed after each rain? Will foggy weather dilute the insecticide to the point that it becomes ineffective?

4. Certain herbs are claimed to deter or repel certain plant pests. You probably know which insects attack your plants year after year. All you need are the herbs, and good-bye pests!

You can grow herbs from cuttings or start them from seeds. Seeds and potted herbs can be bought in garden stores or from seed companies. Potted herbs grow well indoors in sandy soil placed near a sunny window and grow even better in a garden.

Design some experiments to determine the truth of the following:

• Basil repels houseflies. (This experiment can be done inside your house.)

• Chives deter aphids. (Plant the chives close to roses. Rosebuds are a favorite food of aphids.)

• Garlic repels Japanese beetles.

• Mint deters the white cabbage moth.

- Rosemary deters cabbage moths and carrot flies.

- Sage deters the same insects as rosemary.

- Thyme repels the cabbage worm.

Now, try some other herbs and see how they work. You can also try to design a project to answer one or more of the following questions:

- Will a mixture of rosemary, sage, and thyme be a better deterrent to cabbage pests than an organic pesticide?

- Will sprinkling garlic powder on plants work as well as or better than garlic growing near the plants?

- What combinations of herbs work best against the insect pests that inhabit your garden?

- Is it possible to invent a liquid spray made up of several herbs that is effective for a week or more?

PLANTS THAT REPEL INSECTS

Tomatoes and potatoes produce chemicals that prevent some species of insects from digesting their food. These chemicals are released when the tomatoes or potatoes are injured. Can you design experiments to answer these questions:

- Will the juice of a tomato or potato rubbed on tomato leaves and fruit kill tomato hornworms? What is the effect of these juices upon aphids?

- Add tomato or potato juice to any of the homemade insecticides mentioned earlier in this chapter. Are they improved as a result? How well do these juices work against insects that don't feed on tomatoes or potatoes? Are they effective against squash bugs or the corn borer? Do these juices remain effective for longer periods than other organic insecticides you have experimented with?

Cucumber slices have been discovered to repel most cockroaches. What can you discover about the following:

• How does cucumber juice compare with tomato juice as an insect repellent for house plant pests (mealybugs, etc.)? Against garden pests (cabbage flies)?

• What is the effect of a vegetable-juice spray composed of cucumber, tomato, and garlic juices upon insect pests?

• If you live in an apartment, try a combination of bay leaves and cucumber slices as a roach repellent next time cockroaches pay a visit. Bay leaves contain several chemicals that repel cockroaches. Will a combination of bay leaves and cucumber slices repel all roaches instead of most? Experiment with other vegetable juices, such as pepper or carrot juice. Do they repel or attract pests?

Phenols

Recently, it has been discovered that the *replacement leaves* (a second set of leaves grown after the original leaves are eaten) contain larger-than-normal amounts of *phenols* than the original leaves eaten by tent caterpillars. Phenols give insects an upset stomach. Similarly, replacement red oak tree leaves formed to replace leaves eaten by gypsy moths contain more tannins than the original red oak leaves. Tannins also interfered with an insect's digestion of plant material.

You can do these experiments only if your area has had a recent infestation of tent caterpillars or gypsy moths. Their populations go up and down in cycles. For example, in 1979 gypsy moths defoliated 500,000 acres of trees in the U.S. In 1980, the toll was ten times greater, or 5 million acres. In 1981, 12.5 million acres of trees were cleaned of their leaves. In 1982, the gypsy moth population "crashed." They ate the leaves of only 8.2 million acres of trees. Some scientists believe the drop in population was due to steps taken by people, such as spraying with insecticides, using bacteria and viruses that specifically attack the gypsy moth, and releasing sterile male gypsy moths. This last action would result in many of the next year's eggs being sterile. Other experts argue that the increase in phenols and tannins in the leaves of the trees that had been attacked is responsible for fewer insects. Higher than normal levels of tannins and phenols interfere with digestion. Thus, the moths will be smaller, weaker, and less able to reproduce. Perhaps you can resolve this issue through experimentation?

Tent caterpillars. In the early spring, look for tentlike nests on trees where two branches join the main trunk. The larvae tent caterpillars leave the nests and feed on the leaves of the tree during the day. At night they return to their tents.

Gypsy moths lay their eggs in midsummer in strange places including under rocks, cracks in buildings, and on the underside of automobile bumpers. The eggs hatch in May of the next year into brown or black hairy caterpillars with six pairs of red dots on the back. When mature, the caterpillars will measure about 10 cm. Oak, willow, and maple trees are their favorite food but they will eat more than 500 other kinds of leaves.

• Grind up the replacement leaves of either the red oak or willow tree. Allow the ground-up leaves to stay in a small, closed container of water. Use them to prepare a water-based insect spray. Test its effectiveness as a repellent and as an insecticide on local plants and insect pests.

• If the water-based spray shows no positive results, substitute alcohol for water as the solvent. What is the result?

• With the help of a chemist, can you discover what the active ingredients in the leaves are? Can they be combined with other substances to produce a more effective insect repellent or insecticide?

• Test the effectiveness of a leaf solution added to cucumber juice, to garlic juice, etc.

• Fern leaves, particularly those grown early in the year, have been found to contain a number of insect poisons, including tannins, phenol, and glycosides.

Prepare a fern-leaf spray made from ground-up fern leaves. Try a water-based spray made from ground-up fern leaves. Try an alcohol extract if the water spray shows no positive results. What insects are repelled or killed by your fern preparation? How effective is a mixture of fern-leaf juice and other juices such as cucumber?

CITRUS REPELLENTS

Citrus Seeds

Grapefruit seeds taste bitter to both people and the cotton bollworm. The seeds of this citrus fruit contain a group of chemicals called *limonoids*. When cotton plants are sprayed with the juice from grapefruit seeds, bollworms will avoid these leaves unless other food is unavailable. One way to extract the juice of grapefruit seeds is to squeeze them in a garlic press. Collect the juice in a covered jar or bottle.

● Spray grapefruit-seed juice on the leaves of plants growing in your home or garden. If you can, use plants that insects are feeding upon. Do the insects avoid these plants after spraying or do they continue eating?

● Collect and test the juice of other citrus fruit seeds such as lemons, limes, tangerines, or oranges as an insect repellent.

● If any of the citrus-seed juice experiments show positive results, determine how long the juice is effective as an insect repellent.

● Which seed juice repels insects the best?

● Compare the repelling qualities of citrus-seed juice with cucumber juice or garlic juice.

● Does citrus-seed juice work better as an insect repellent if the spray is made with alcohol rather than with water?

● Extract the juice from grapefruit seeds. Add the juice to previously made organic pesticides. Does the combined insecticide now repel more insect pests?

● Does feeding on the juice kill any species of insect? If so, which?

● You can work on one question only or try to answer more than one.

Citrus Skin

If you grate the skins of citrus fruits and then crush them in a garlic press, an oil will be released. Collect this oil, because it has been found to paralyze and then kill houseflies, stable flies, and paper wasps. (These insects annoy people in a number of ways.) Don't overlook this oil in the search for a replacement for synthetic chemical insecticides. The insect-killing properties of citrus oil (made from orange peel) was first discovered in a hand cleaner named Dirt Squad. This happened when the hand cleaner was accidentally sprayed on houseflies.

- Is the oil from a citrus fruit skin identical to the juice in seeds of the fruit? Do both produce the same effects?

- Will the oil from orange or lemon skins kill tomato cutworms or aphids?

- What is the effect of citrus-fruit skin oil on mealybugs and white flies?

- What are the repelling and killing effects of a mixture of orange-skin and orange-seed oil on various insect pests?

- Will an orange peel "collar" around a plant stem protect that plant from insect pests that crawl up the stem from the soil? You could put "collars" made from the skins around the stems of tomato plants. Does the "collar" repel tomato cutworms or slugs? In order to have adequate controls, place "collars" of the same size around your control plants. These collars should be made of cardboard or plastic. Remember, the plants must be as alike as possible in both the experimental and control groups.

- How does a citrus-peel "collar" compare with a citrus-seed oil spray? Which deters or kills more pests? Which remains effective longer? Do the collars have to be replaced after each rain?

- Experiment with other combinations of citrus plant parts.

WHAT THE WORLD NEEDS

The world awaits an organic, wide-spectrum insecticide, one that will kill many insect pests without harming plants, birds, animals, beneficial insects, or people. You may be the one to develop that insecticide. Go for it!

9

EXPERIMENTS— ODDS AND ENDS

It is possible that you have arrived at this chapter without having selected a project. Here you will find a number of experiments and projects that didn't fit neatly into the other chapters. They are arranged in no particular sequence of importance or degree of difficulty. Look them over, and maybe one of them will turn out to be for you.

MAGNETIC AND ELECTROMAGNETIC FIELDS

Magnets and electromagnets are important aspects of our lives. Without them there would be no electricity or wonderful electrical devices such as the TV set, Walkman™, and the telephone. The earth itself is one giant magnet that produces a magnetic field which is invisible and affects iron and other metals. Do magnetic and electromagnetic fields affect the germination, growth, and development of seeds and plants (Fig. 9.1)? Here is an experiment to try for "starters":

1. Divide 100 radish seeds into two groups. Label one group "Exp" and label the other "Con."

2. Place each group in a separate germination chamber (see chapter 3). Germinate your "Con" group under normal conditions.

3. Germinate the "Exp" group under identical conditions except place the growth chamber (a) between the ends of a horseshoe

Figure 9.1. Will the magnetic field of this horseshoe magnet affect the growth of these seedlings' roots?

magnet; or (b) between the north poles of two bar magnets. Record the distance from the poles to the center of the germination chamber (repeat this experiment using the south poles); or (c) 5 cm from the end of an electromagnet of known strength. Maintain the experimental and control conditions until 75% or more of the seeds have germinated.

Which group has a higher germination rate? Which group germinates sooner? Are the magnetized seedlings normal in appearance, size, and color? Do your results differ if the magnets are above and beneath the germinating seeds instead of being in a horizontal plane? If you wish, continue to keep the experiment going until the plants mature.

SOUND AND MUSIC

A popular experiment has been to expose plants to sound in order to determine if plants "hear" (see Fig. 9.2). Some people claim that plants grow better if exposed to soft, soothing music than to loud rock or disco music. Other people claim that plants that are spoken to nicely and politely grow better than plants that are yelled at in a rough, loud voice.

When doing experiments with sound and music, remember that a negative answer is just as meaningful as a positive conclusion. Not all experiments end up with dramatic new findings; in fact, most don't! Second, it is hard not to give some slight advantage to the experimental plants because all of us would consciously or subconsciously want to have our experiment end in a positive way. (In this case, we might want our plants to respond to music or to our voice.) What usually happens is that the experimental plants get slightly better care—a little bit more water or light. One procedure used to keep professional experiments free of unconscious bias is to use two groups of scientists. One group works with only the experimental group and the other with only the controls. Neither knows what the experiment is designed to discover or that the other group exists, until after the experiment is concluded. Both groups work inde-

Figure 9.2. Will lettuce in this pot grow better if you play love songs to it?

pendently and carry out the experimental procedures called for by the scientist in charge. As a good research scientist, you must try to treat the experimental and the control plants identically except *for one factor* and recognize that negative results have value.

Here, then, are ideas for experiments using sound and music. You can, of course, think up experiments of your own.

1. From the same seed package, prepare two groups of 50 seeds (lettuce) for germination under the same conditions of temperature, darkness, etc. Germinate one group in a little used room that is far from noise. Germinate the other group in another room, where a radio is playing continually.

Which group germinated sooner? Which group had the higher germination percentage?

2. Select ten seedlings from each group that are as alike as possible. Pot them using the same mix. Continue the experiment for three months. Caution! Keep all conditions except for sound as identical as you can for both groups of plants.

Which group grew larger? Faster? Had more leaves? More flowers? Compare the two groups in other ways such as the dry weight of the ten plants in each group.

3. Repeat the above experiments (a) using classical and popular music; (b) telling one group that you love them every morning, and screaming at the other group every morning that you hate them; (c) trying different plant seeds. Are your results always the same regardless of the species of plant?

If you get negative results from these projects, you might want to experiment with high-frequency sounds out of the range audible to humans.

Ultrasonics

Sound waves are produced by vibrations that travel through air or some other medium. Our ears can hear waves that range from 16 to 20,000 vibrations per second. Other animals can hear higher vibrations called *ultrasonic waves*. Bats, for example, use ultrasonic waves to locate objects in the dark. Dog trainers use ultrasonic whistles to train dogs. Perhaps plants "hear" ultrasonic sounds? Experiment and find out.

• Purchase an ultrasonic dog whistle in a pet shop. Set up two groups of identical seeds or seedlings. Expose one group to the normal sounds in your house or apartment. Twice a day (morning and night), sound the ultrasonic whistle for five minutes close to the other group. Make sure this group is isolated from normal sounds.

• What differences between the groups can you observe in two weeks? In one month? In three months?

• Ultrasonics has been used to kill bacteria. Will ultrasonic sounds kill or shake out insects such as aphids or tomato hornworms? When these pests are feeding on your plants, try ultrasonics. Be sure to keep the ultrasonic sound going for the same amount of time (15 minutes). Do this several times a day. Don't give up, even if nothing exciting happens the first time. Keep up the experimental procedure for at least two weeks.

• Certain electronic devices are supposed to repel roaches and other pests. These devices emit ultrasonic sounds. You could direct their sounds at plants to see if the vibrations have any effect on plant pests.

• Ultrasonics is also used to clean clothes (by shaking out the dirt), sterilize food, homogenize milk (evenly distributing the fatty cream throughout the watery milk), and perform bloodless nerve and brain surgery. Do any of these uses of ultrasonics suggest ideas for plant experiments?

PLANT COMMUNICATION

Plants may not talk to one another using words, but they may release chemicals into the air that do enable them to communicate. Scientific evidence supporting this phenomenon has come from two scientists, Gorden Orians and David Rhoads, at the University of Washington. Although the chemicals are as yet unidentified, they are thought to be *pheromones* (chemicals that carry information), the same group of substances that insects use to attract members of the opposite sex. A good deal is known about insect pheromones, but the idea of plant pheromones is new and exciting.

The two scientists experimented with willow trees for four years before announcing the results of their work. They found that willow trees being destroyed by tent caterpillars and web worms sent messages to nearby willow trees that were not under attack. These healthy trees *changed the chemical nature of their leaves so that the leaves were not tasty to the insect pests!*

Since plant scientists are having trouble identifying plant pheromones, you probably should attempt this type of work only in collaboration with an experienced plant researcher. However, there is plenty of room for other important projects and experiments that can add to our knowledge of plant pheromones.

• Do trees other than willows also communicate? In the previous chapter, you read about red oak trees that made chemical changes in their leaves after being eaten by gypsy moths. Survey the red oaks in your area. If you find some that gypsy moths are feeding on, look for nearby red oaks that are free of the moths. These nearby trees may have been "warned" by the attacked red oaks. It might be worthwhile to chemically analyze leaves from attacked red oaks and those that seem to be immune to gypsy moths. To do this properly, you will need the assistance of a chemist with the appropriate laboratory equipment.

• Do flowers and vegetables attacked by insect pests also "warn" their relatives? Find out by making surveys in the spring and summer when insects are feeding. When you find a plant species being eaten by insects, look for other, nearby plants of the same species that are insect-free.

"A" IS FOR AUXIN

Auxins, gibberellins, and cytokinins are the three major plant hormones that regulate growth. Auxins were the first hormones to be discovered in the *apex* (growing tip) of roots and stems. It was found that if you remove the tip of a stem that is growing rapidly, the growth dramatically slows down and then stops. Early researchers removed the tips and placed them on blocks of gelatin. A few hours later, the gelatin block was placed on the cut stump. The researchers were amazed to discover that growth resumed. You can try this experiment for yourself. The hormone was analyzed and named *indole-3-acetic acid* or IAA. As the concentration of IAA is increased, growth is speeded up. However, if the concentration is further increased, growth slows down and finally stops.

1. Prepare various concentrations of IAA, from 0.0001 g per liter to 0.1 g per liter. (IAA is available from scientific supply companies.)

2. Cut the tips off of pea plants or any plants with rapidly growing stems; place the plants, not the tips, in petri dishes, and add the IAA solutions. Observe at intervals for 72 hours.

At what concentration does the most growth take place? At what concentration does growth slow down? Stop?

3. Repeat this experiment, applying the IAA solutions to the cut ends of bean plants. What are your observations? Your conclusions?

Auxins have been observed to have many and varied effects upon plants. Following are ideas to ponder, questions to answer, potential projects:

• Adding auxins to stored potatoes prevents buds from forming. Can onions sprayed with auxins be stored for longer periods of time than onions not sprayed?

• Leaves fall as soon as the *petiole* (the part that attaches the leaf blade to the stem) produces too little auxin. Spraying leaves with certain auxin combinations will prevent this. Such sprays are commercially available. Will this work on house plants that normally drop their leaves in the winter time?

• By applying a synthetic auxin (alpha-naphthaleneacetic acid), pineapple and litchi nut plants have been made to flower. If you apply alpha-naphthaleneacetic acid to the ovary of tomato flowers, fruits will form without pollination occurring. These fruits are seedless. Try to produce seedless tomatoes or cucumbers.

• 2,4-D (2,4-dichlorophenoxyacetic acid) kills dandelions and all other broad-leaved weeds growing in cornfields. The narrow-leaved corn plants are unaffected. How long does 2,4-D remain effective? What is the lowest effective concentration of 2,4-D?

"C" IS FOR COCONUT

Coconut milk contains several auxins and nutrients that stimulate the growth of some plants. Drain the milk from a coconut and store the milk in the refrigerator for experimental purposes. Even under refrigeration it will spoil if kept too long. After a week, check your coconut milk by smelling it. If bacteria are fermenting it, your nose will let you know.

• What is the effect of coconut juice upon the germination rate of radish seeds? In this experiment, use two controls: cow's milk and water.

• Do tomato seedlings watered with coconut milk grow faster and/or taller than those given water?

• What is the effect on the peas if coconut milk is sprayed on green pea flowers?

• Will lettuce watered with coconut milk be tastier or have larger leaves than lettuce plants that get only water?

"C" IS FOR CYTOKININS

Cytokinins are plant substances that promote the formation of buds. They also slow down the yellowing and death of leaves removed from growing plants. Kinetin is a synthetic chemical compound that produces the same effects as natural cytokinins.

Here are a few experiments you might decide to perform:

• Take leaf cuttings from house plants such as coleus or African violets. Be sure the petiole is left on the blade. Root the petioles in sand containing kinetin solutions in varying concentrations (1 part per 100,000 of solution to 1 part per 10 of solution). Use pure water as your control. Which concentration of kinetin causes roots and stems to form the fastest?

• Will roots of plants grow stems and leaves under the influence of kinetin? Work with 6-cm pieces of roots of morning-glory, carrot, African violet, tomato, etc. Use a number of different concentrations and pure water as the control.

• Repeat the two previous experiments but include a solution that is a mixture of kinetin and coconut milk.

"G" IS FOR GIBBERELLINS

About 100 years ago, Japanese rice farmers noticed that some of their rice seedlings grew exceptionally tall. However, the plants never flowered or lived a normal length of time. In 1955 one of the substances responsible for such growth was discovered: gibberellic acid. Scientists learned that the most dramatic effect of gibberellic acid was to lengthen plant stems. In a few cases, it reduced leaf size. Today, one of gibberellic acid's main uses is to get "long-day" plants to flower quickly (see chapter 4 and Fig. 9.3).

• Water five 1-sq-m areas of grass with a solution of dilute gibberellic acid (10%, 1%, 0.1%, 0.01%, 0.001%). Observe the rate of growth and the height, health, resistance to fungi, resistance to insects, etc., in each plot. Compare with data from plots that receive only water.

• Prepare the same solutions as above. How do the solutions of gibberellic acid affect the growth, taste, and height of celery plants?

• Gibberellic acid is available as a paste. Apply the paste to seedless grape clusters, regular grape clusters, tomato flowers, Jerusalem cherry flowers, and apple blossoms. Compare resulting plants with controls grown without the use of gibberellic acid paste.

• Expose the seeds of normal and dwarf peas to gibberellic acid. Treat the seedlings with a spray of 0.01% gibberellic acid. What can you conclude is the effect of gibberellic acid? Repeat using dwarf and normal-size corn seeds.

Figure 9.3. What will gibberellic acid do to these petunia flowers?

● Apply gibberellic acid to dwarf and normal-size flowers such as daisy or primrose. What are your observations and conclusions?

Gibberellins make plants grow tall while other growth regulators such as cyocel and N-dimethylamino succinic acid cause plants to be dwarfs. Cyocel prevents plants from making gibberellins. Try any of the gibberellic acid experiments with either dwarf-producing growth substance.

LUNAR INFLUENCES

The phases of the moon supposedly affect the lives of people and animals. Quite possibly plants are affected by the position of the moon in relation to the sun and the earth. Unfortunately, experiments that test lunar effects are difficult to properly set up. How can you be sure that whatever effect you discover is really due to the new or full moon? It's easy to apply a paste of gibberellic acid to ten grapes on a grapevine and compare these grapes with identical grapes right next to them that did not have the paste applied. It's less easy to detect "lunar" influences. You certainly should try, however, and here are a few ideas to work with:

• Do more seeds germinate if started at the new moon phase than at the full moon phase? In order for this experiment to be controlled, *all* other factors—water, temperature, light, etc.—must be the same for both groups.

• Is plant growth more rapid at one phase of the moon than at others? You could measure the average length that stems grow or the number of new leaves as indications of growth. One way to measure stem growth is to put an ink or dye mark at the end of a growing tip as your reference point. Measure daily.

MOTION EXPERIMENTS

You can explore the effects of motion on plant growth in many ways.

If you have a grandfather's clock or any clock with a pendulum, find out whether continual motion has any effect upon seedlings. Place some lettuce seeds on moist blotting paper and seal the seeds and moist paper inside a plastic bag. Fasten the plastic bag to the pendulum with tape. Examine the seeds daily. Compare their growth and development to that of nonmoving controls grown in the same room as close as possible to the moving seeds.

Another possibility is to use tomato seeds that have orbited the earth aboard the space shuttle. In April 1984, 12.5 million tomato seeds were placed aboard the shuttle. These seeds are for use by school-age youngsters in germination and growth experiments. The seeds will be available starting in 1985. Your science teacher will be able to provide you with information about getting hold of some of these unusual seeds. What experiment would you like to do with tomato seeds that had orbited the earth for 12 months?

ACID RAIN

Normal rain tends to be acidic, with a pH between 5.6 and 5.7 (neutral pH = 7). The reason is that the carbon dioxide combines with the moisture in the atmosphere to form carbonic acid, a weak acid. This is the same acid that causes sodas to taste sour. If the pH is below 5.6, the rain is said to be *acidic*. In Wheeling, West Virginia, the pH of a rain storm in 1978 was less than 2.0. These rain drops would taste like vinegar!

Other types of acid precipitation occur besides rain, including acid fog, acid snow, acid hail, and acid sleet. The term "acid precipitation" would be better to use instead of acid rain, since so many types of acid precipitation occur. Since pH is so important to understanding acid rain, you may wish to reread the section on pH in chapter 3.

Acid precipitation is caused by the oxides of sulfur and nitrogen released into the atmosphere. These come from electrical generating plants, industrial factories, cars, and trucks. When these oxides combine with moisture in the atmosphere, they form sulfuric acid and nitric acid. As our need for electricity, new products, and transportation grows, so does the problem of acid rain.

The effect of acid rain on plants is poorly understood and is being studied. In fir trees, the first visible effects of acid rain are yellow needles and drooping branches. Early investigations concluded that acid rain damages the surface of leaves, interfering with transpiration and lowering the rate of photosynthesis. Acid precipitation also damages the root system, and lowers the germination rate of seeds. However, some scientists now believe that these harmful effects are not due to acid rain but can be attributed to summer droughts or high levels of ozone in the atmosphere. These differing points of view indicate a new, wide-open area for experimentation.

Before you can begin to experiment, you will have to prepare an acid solution that mimics acid rain. Since acids are extremely destructive to living tissues, make sure you follow these safety precautions:

- Wear safety goggles at all times.

- Always pour acid into water, never the other way around.

- Wear an apron or old clothing with long sleeves to protect against splattering.

- If acid does get on your skin, wash it off with lots of water. Neutralize any remaining acid by applying sodium bicarbonate.

• Wash your hands thoroughly after working with acids.

Most experts feel that the harmful effects of acid rain to lakes, soil, and rock are due to pH rather than to a particular acid. Damage to plants may result from other causes. Several scientists report that some plants are helped if the acid rain contains dilute nitric acid, since this acid adds nitrates to the soil. Likewise, soil lacking sufficient sulfur for plant growth can actually be improved by sulfuric acid rain. *Too much* sulfur or nitrogen can be harmful to plants.

Here are formulas for making sulfuric acid (H_2SO_4) rain of different strengths and nitric acid (HNO_3) rain.

Sulfuric Acid

1. Start with spring water, which can be purchased in the supermarket. Check its pH with hydrion paper; it should be 6. You also will need a 10% solution of sulfuric acid.

2. Add 10 drops of 10% sulfuric acid solution to 1000 ml of spring water. Stir thoroughly. The pH is 5.

3. For a pH of 4, use 1000 ml spring water and add 30 drops 10% sulfuric acid. Stir well!

4. Adding 50 drops of 10% sulfuric acid to 1000 ml spring water yields a solution whose pH is 3. Stir well.

5. Adding 60 drops 10% sulfuric acid to 1000 ml spring water yields a pH 2 solution. Stir well.

Nitric Acid

1. Ask your teacher or a knowledgeable, experienced adult to prepare a 0.1 molar (M) solution of nitric acid by adding 3.25 ml concentrated nitric acid to 500 ml spring water.

2. Add 2 ml of the 0.1 M nitric acid solution to 198 ml spring water. The pH of the resulting solution will be 2. The solution will also be 0.01 M nitric acid. Use this to prepare a nitric acid solution of pH 3.

3. Add 2 ml of the 0.01 M nitric acid solution to 198 ml spring

water. The resulting solution's pH is 3, and it is 0.001 M HNO₃.

4. To get a pH of 4, add 2 ml of the 0.001 solution to 198 ml spring water.

5. Making a nitric acid solution of pH 5 or 6 is easy. Just make another dilution or two.

Now that you have your laboratory "acid rain," you may want to perform the following experiment.

Root Growth

What is the effect of acid rain upon the growth of roots on coleus leaf cuttings?

1. Prepare jars of distilled water, tap water, spring water, nitric acid water, sulfuric acid water with pH values between 2 and 5. Your control is the water closest to the neutral pH (7).

2. Use hydrion paper to determine and then record the pH of the liquid in each jar. Label and cover the jars.

3. Prepare three coleus leaf cuttings (see chapter 5) for each jar of water you plan to use. All cuttings should be similar in size, state of health, and age. Pot the cuttings in small jars using sand or vermiculite as the rooting material. Your containers and the amount of rooting material must be the same in all cases.

4. Place all the rooted cuttings so that all are at the same temperature, and get equal intensities and amounts of light.

5. Keep records of the amount of water added to the cuttings.

6. Each day, check for the growth of roots.

7. When roots start to form, use a metric ruler to determine the length of the roots. Measure from the tip of the longest root to the point where the root enters the stem. Do this daily for five days. Be sure to keep accurate records.

In which solution did the roots grow the best? The poorest? Prepare a graph of the root growth for each pH used. Plot the daily growth over the five-day period against pH.

Spin-off Experiments

• Use begonias and other plants rather than coleus plants.

• What is the effect of various pH solutions made by mixing nitric acid and sulfuric acid water?

• What is the effect of acid rain water on the germination rate of lettuce seeds? On the germination of other seeds?

• What is the effect of different pH raindrops upon the surface of African violet leaves? On the leaves of other plants?

• At what pH is the transpiration rate of a geranium plant the slowest? The fastest?

• What is the highest pH that will kill a dwarf tomato plant? In which pH range do tomato seedlings grow the fastest? The slowest? Produce the tallest stems? The most leaves?

• Add a root-forming substance (IAA) to the water in all the coleus leaf cutting containers. What is the effect of acid rain water on the action of IAA?

BE AN ACID RAIN WATCHER

In many countries, scientists are collecting data on acid rain. They set up collecting stations; after each rain or snow storm, they determine and record the pH of the precipitation. They also record the direction of each storm. This can lead to identification of the pollution source. Once the source is known, elimination of the pollution becomes possible. Of course, effecting change is extremely difficult for various social, economic, and political reasons.

If you would like to become an acid rain surveyor, here is a way to get started.

1. Set up acid-rain- and acid-snow-collecting stations (plastic or glass containers) on the roof of your apartment building or house, or in your yard or garden. Set up two or three stations and use an average pH for your observations. This will help you create accurate records.

2. Record the dates of storms, and the direction from which

they come. You can learn the direction of the winds and the storm's path by watching the weather segment of the local TV news programs, listening to the radio, reading your newspaper, or watching the in-depth weather coverage on some cable TV channels. You can also use a weather meter (available from most gardening supply companies). Such a meter can measure wind speed, wind direction, temperature, and the day's rainfall, and they are fairly inexpensive. Ringer Research, 8660 Flying Cloud Drive, Eden Prairie, MN 55344, sells one that is durable and reasonable in price.

3. After each rain or snow storm, measure the pH of the precipitation in the collecting containers. You will be making many measurements over many months, and hydrion paper is neither the fastest nor best way to determine pH. Purchase an accurate pH meter (sold by gardening supply companies). The main advantage of the pH meter is that you can determine the pH quickly. Also, the meter can be used over and over again. Meters can be purchased from Ringer Research (address above), from Park Seed Co., S.C. Hwy. 254 N., Greenwood, SC 29646, and from other companies.

4. Keep accurate records over a three-month period or longer.

Here are some questions you can try to answer:

● From which direction do the worst acid rainfalls come? How can you account for this observation?

● During which season of the year is the acid precipitation the worst? Can you discover why this is so?

● What is the average pH of all the precipitation you have recorded? How does it compare with that in other parts of the United States?

● Using a soil test kit (sold by gardening supply firms and chemical supply companies), analyze the amount of nitrogen, phosphorus, and other minerals in the soil at weekly intervals. Analyze the soil near your collecting containers. Record the results. Which mineral is lost in the largest amount from the soil due to acid rain action?

● Set up a network of acid rain observers among your friends.

Prepare tables, maps, and graphs of the acid rainfall in your town, city, or county.

• A few years ago, some observers in New York State reported that acid snow contained two parts of nitric acid to one part of sulfuric acid. In summer rains, they reported two parts of sulfuric acid to one part of nitric acid. Determine, by chemical means, what acid or acids are in the acid precipitation in your locale. Does the composition of the acid precipitation that you collect vary from season to season?

• If your state has established monitoring stations for sampling acid rain, contact the people who run them. Perhaps your data could be part of their official survey. The U.S. Geological Survey coordinates the sampling done by the various states. You can write or call this federal agency or the U.S. Environmental Protection Agency in Washington, D.C., for assistance or for information regarding acid rain surveys.

10 REPORTING YOUR WORK

After completing your experiment or project, you will want to make other researchers aware of what you did and what your results were. This is always done by means of a written report. If you wish to enter your work in a science fair, you will need a visual display as well (Fig. 10.1).

Figure 10.1. The written report is an important part of this science fair entry.

THE WRITTEN REPORT

Your science research paper should follow a prescribed order. Sometimes that order is prescribed by a science competition, such as the Westinghouse Science Talent Search, which has its own format. The scheme that most beginning and professional researchers use for reporting their results will be explained in detail. Here are the major parts of a science research paper arranged in that order:

Title Page
Abstract
Introduction
Procedure
Observations
Conclusions
Recommendations
Bibliography

Title Page

The title of the research paper should appear in the center of the page. Type your name and address on the title page. You might wish to include your school's name and address. If this report is for a particular teacher, put that teacher's name and your class subject on the title page. The last thing that you need on this page is the date. Look at the sample that follows to see exactly what a title page should look like.

THE EFFECT OF GIBBERELLIC ACID UPON
TAP ROOT LENGTH IN DAUCUS CAROTA

James Smith   Mr. R. Jones
5 Center St.   Kennedy HS
Adamsville, NJ   5/23/86

Abstract

This is a brief summary of what you did: the problem, experimental plan, observations, conclusions, and recommendations. The abstract should be approximately 150 words long, but if you write less than that, don't worry: the shorter, the better. Even though the abstract comes first, write it after finishing the rest of your report. Leaving it for last is wise because you can copy or rephrase some work directly from the main report. The abstract is the first page after the title page. Label this page ABSTRACT (use capital letters).

Introduction

Label your next page INTRODUCTION. In this section, you give the reader some of the background information that you were able to find. Include the important experimental results of earlier researchers. You might want to tell the reader why this problem is important or why you decided to research this area of plant science. One function of this section is to get the attention of the reader and make him or her want to read the rest of your report. This section should be between one and two double-spaced typed pages.

Procedure

Under this heading, tell the reader what you did and how you did it. Be precise because the reader may want to repeat your experiment. No one can repeat your experiment unless they know exactly what you did. For example:

DO NOT WRITE: "I treated the seeds with a dilute solution of gibberellic acid. Then, I incubated them for a while."

DO WRITE: "I moistened the ten dwarf tomato seeds with 15 ml of a 0.1% solution of gibberellic acid. I incubated the seeds inside a sealed petri dish for 72 hours at a constant temperature of 23° C."

Outline your experimental plan step by step in a similar manner.

Observations

Observations are also called "results." No matter what you call them, observations—or results—are the information you discovered when you did your experiment or project. Your results (data) are presented in written form. Once again, your writing must be as precise as possible. Writing "I observed that the radish seedlings roots grew best when I used a yellowish-orange filter over the light" is imprecise. There is too much the reader doesn't know. A more precise statement would be, "Roots of radish seedlings grew the longest when I used a filter that allowed yellowish orange light with a wavelength of 6300 Å (an Angstrom unit is equal to 10^{-10} m) to reach the seedlings. All filters were placed in front of 100-watt bulbs placed 50 cm from the seedlings." This statement says much more!

To make your data even more meaningful, include tables, graphs, and photographs in this section of your report.

Tables

When your results are displayed in a table, they are organized logically and look neat. Tables allow for easy comparisons of data. In the table that follows, you can easily compare the effect of red light with that of any other color you wish. Looking at a table makes understanding what was written easy. Always give titles (EFFECT OF LIGHT UPON ROOT LENGTH OF RADISH SEEDLINGS). Tables have several columns, and each column is given a title (Filter Color, Wavelength, Length of Roots, Day 2, Day 4, Day 6, Day 8, Day 10).

Table 10.1

EFFECT OF LIGHT UPON ROOT LENGTH OF RADISH SEEDLINGS

Filter Color	Wavelength (Å)	Length of Roots (mm)				
		Day 2	Day 4	Day 6	Day 8	Day 10
Red-orange	6900	0.5	2.5	4	7	9
Yellow-orange	6300	3	6	9.5	13	16
Yellow-green	5500	0	2	3.5	5	8
Green	5100	0	2	3	4	5
Blue	4700	1	4	7	10	13
Purple	4000	0.5	2	2.5	3	3.5

Graphs

Graphs help you and the reader to draw conclusions from your data. They also permit you to formulate generalizations. The data in Table 10.1 was our source of data for the graphs that follow. We can draw the conclusion that light of 6300 Å led to maximum root growth. Another conclusion is that two wavelengths produce good growth; 6300 Å and 4700 Å. We can generalize from the graphs that radish seedling roots will grow in all wavelengths of visible light.

There are several kinds of graphs. For our purposes, we will consider only *bar* and *line* graphs. Pie graphs are useful tools for economists and social scientists, but are rarely used by plant scientists. Pie graphs are excellent for showing how things are divided into their many parts, for example, what percentage of students who read this book did plant experiments or projects (Fig. 10.2).

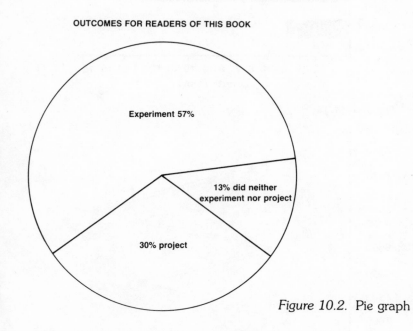

OUTCOMES FOR READERS OF THIS BOOK

Experiment 57%

13% did neither experiment nor project

30% project

Figure 10.2. Pie graph

All graphs must have a title. Both bar and line graphs have a horizontal and a vertical axis. Each axis is labeled and divided into equal parts.

Bar graphs are diagrams that use bars rather than a line to represent the change in a variable factor (growth) in an experiment. Bar graphs reveal relationships that exist between data sets. You can easily see which wavelength produced the greatest amount of root growth, the second greatest amount, and so on (Fig. 10.3).

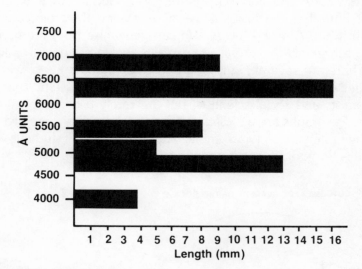

Figure 10.3. Bar graph

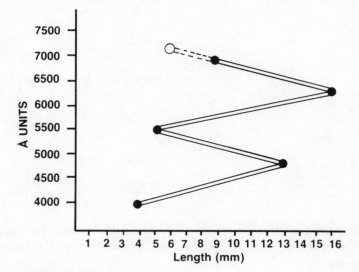

Figure 10.4. Line graph

Line graphs are made by a set of points drawn as a line or a curve to represent the change in a variable (growth). Although they closely resemble bar graphs, they are more useful. First, you can estimate the data between two plotted points. For example, you can estimate the root growth for 4300 Å light even though no point exists for that wavelength on the graph. Second, you can extend line graphs in order to predict data not on the graph. The dotted line predicts the root growth for 7200 Å light (see Fig. 10.4).

Photographs

Photographs say a lot and serve to save you words in your written report. They also verify your data and make your report more appealing to read. Your pictures should be clear and sharp. They should clearly show the viewer what you want him to see. Fuzzy, very dark, or very light pictures should not be used. If you do use photographs, plan beforehand and take them as your experiment or project progresses. Often it is impossible to take a key photo *after* the experiment is over. Color and black-and-white photographs have different advantages, and which you use will depend on the nature of your project. Remember, however, that most magazines (if you're thinking of trying to publish your work) ask for black-and-white pictures.

As TV cameras and VCR's become more inexpensive and available, recording your results on a segment of videotape is also going to be a real possibility for making your research report visual in an impressive manner.

Conclusions

In this section, you bring together all the information you have collected about the problem investigated. Your information includes references to what you have read on the topic and the data you gathered during the course of the experiment. Here, you logically answer the question raised by your work. A plant experiment is really a question asked of a plant. The plant can't give you an oral or written answer, but does give you an answer in terms of growth, development, or reproduction. A good idea is to go back to your hypothesis. Do the data agree with your hypothesis? If so, your conclusion is that your hypothesis is correct. If not, then you must report negative results. Remember, negative results are equally as important as positive results! Don't be ashamed or afraid to come to a negative conclusion concerning your hypothesis.

Recommendations

These are an outgrowth of your conclusions. Some science writers lump recommendations together with conclusions. If this seems best for you, don't be afraid to do it. Usually, your recommendations suggest areas for others to research. Sometimes, your recommendation may be that something not be done. For example, if you work with a particular herb as an insect repellent and find that it has no effect on a dozen different insects, your recommendation should be that no further investigations using this herb as an insect repellent be tried.

Bibliography

This is the last part of your research paper. In it you list all of the readings that helped you get background information in order to do your experiment or project. Include books, articles, and even newspaper stories in your bibliography. Each type of source is listed in its own special way. List entries in alphabetical order by authors' last names. Different types of references have different formats. If you had done a project on plants and acid rain, your bibliography might include the following:

Faber, Harold. "Project to Check Adirondack Lakes." *The New York Times,* November 20, 1983, pp 7–8.

Likens, G. E. "Acid Precipitation." *Chemical & Engineering News,* November 22, 1976, pp 29–44.

Perley, Howard. *Acid Rain: The Devastating Impact on North America.* New York: McGraw-Hill, 1982.

The first is for a newspaper, the second for a periodical, the last for a book. Detailed bibliographic formats can be found in most English texts and also in reference books such as the University of Chicago Press's *A Manual of Style.*

BIBLIOGRAPHY

Abraham, G., and K. Abraham. *Organic Gardening Under Glass.* Emmaus, Pennsylvania: Rodale Press, 1975.

Bartholomew, M. *Square Foot Gardening.* Emmaus, Pennsylvania: Rodale Press, 1981.

Beller, J. *So You Want to Do a Science Project!* New York: Arco Publishing, 1982.

Biological Control of Plant Pests. Brooklyn, New York: Brooklyn Botanical Garden, 1976.

Copeland, L. *Principles of Seed Science and Technology.* Minneapolis, Minnesota: Burgess Publishing, 1976.

Dale, J. E. *The Growth of Leaves.* Baltimore, Maryland: University Park Press, 1982.

Denisen, E. *Principles of Horticulture.* New York: Macmillan, 1979.

Faust, Joan Lee. *The New York Times Book of House Plants.* New York: The New York Times Book Co., 1973.

Free, M., and M. J. Dietz. *All About House Plants.* Garden City, New York: Doubleday, 1979.

Hartmann, H., and D. E. Kester. *Plant Propagation.* Englewood Cliffs, New Jersey: Prentice-Hall, 1975.

Kramer, J. *Indoor Trees.* New York: Hawthorn, 1975.

Nichols, Richard. *Beginning Hydroponics.* Philadelphia, Pennsylvania: Running Press, 1977.

Rayle, D., and H. Wedberg. *Botany: A Human Concern.* Philadelphia, Pennsylvania: Sauders College, 1980.

Rice, L. W. and R. Rice. *Practical Horticulture.* Philadelphia, Pennsylvania: Sauders College, 1980.

Saunby, T. *Soilless Culture.* Levittown, New York: Transatlantic Arts, 1972.

Shepard, J. F. "The Regeneration of Potato Plants From Leaf-Cell Protoplasts." *Scientific American,* May, 1982, pp 154–166.

Skinner, B. J. *Use and Misuse of the Earth's Surface.* Los Altos, California: William Kaufmann, 1981.

Wright, R. C. *The Complete Book on Plant Propagation.* New York: Macmillan, 1975.

INDEX